Werewolves

Encounters with the Black Dog

Joedy Cook

Werewolves : Encounters with the Black Dog© 2016 by Joedy Cook . All rights reserved. No part of this book may be used or reproduced in any manner whatever., Including internet usage , Without written permission from me (Joedy Cook).

First Edition

First Printing. 2016

Book design by Joedy Cook

Interior photos by contributors

Library of Congress

ISBN-13: 978-1536827262

ISBN-10: 1536827266

Printed in the United States of America

DEDICATION

To my wife Pamela for supporting me for the last sixteen years. And to all my friends in the field who worked as hard as I have looking for the truth.

TABLE of CONTENT

Foreword– Vic Cundiff   Pg-7

Introduction- WHY WEREWOLVES AND DOGMAN CAN EXIST   Pg-9

Chapter One - Joedy Cook
Historical review of the Black Dog Pg-13

Chapter Two– Adam E. Davis
Werewolf of Germantown   Pg-29

Chapter Three– Anthony J. Chaney
Beast in the field   Pg– 39

Chapter Four– David Leidy
Bigfoot Dogman and Human Pg-50

Chapter Five– Linda Godfrey
Metal mouthed Dogman of Clare Pg-69

Chapter Six– Brian Seech
The Shenago valley Werewolf Pg-75

Chapter Seven – Ronald L. Murphy
Encounters with a Dogman   Pg-85

Chapter Eight – Danielle Steadman
Wolfman   Pg –94

Chapter Nine – Donna Fink
Something Unexplainable   Pg-113

Chapter Ten-KBRO
What an I seeing   Pg-123

Chapter Eleven-Mike Lawrence
Their watching me   Pg-126

Chapter Twelve – Mike
It's tracking me   Pg-132

Chapter Thirteen – Dianne Beeson
Naked Dogman   Pg-135

Chapter Fourteen– Vic Cundiff
Werewolf of Arkansas   Pg-137

Chapter Fifteen– NADP
Werewolf sightings in the UK   Pg-141

Chapter Sixteen– NADP
Beast of the land between the lakes   Pg-149

Chapter Seventeen-NADP
Werewolf in Erlanger, Ky   Pg-164

Chapter Eighteen-Vic Cundiff/NADP
The Black Dog   Pg-166

Foreword

By
~Vic Cundiff~

They're out There!

It might be hard to believe, but over 40% of North America is still covered by forests and other wooded lands. In these wild places, eyewitnesses report seeing large, unclassified creatures which are called, by many, "Dogmen." "What's a Dogman?" you ask. Well, that's a difficult question to ask. The reports, that come in, from eyewitnesses describe fierce-looking creatures that share some traits but differ in several ways. While Dogmen can have variances in their appearance, all Dogmen seem to have heads that resemble a K9 or a baboon, large, sharp teeth, and a tendency to use bipedal locomotion when it suits them.

Most eyewitnesses, when they encounter a Dogman, are left unharmed. On some, rare occasions, eyewitnesses aren't so lucky. In almost every encounter, it would seem, the Dogman's sole intention is to scare the eyewitness to within an inch of their life. Once that's been accomplished, the Dogman usually moves on.

Just because you don't venture into the woods, don't think that precludes the chance of you having your own encounter, with a Dogman. People have had encounters with them, well within the city limits of some major cities, such as Chicago and Cincinnati. Being in such a developed area does greatly reduce your chances of having an encounter, however.

The next time you're taking out the trash, late at night and you hear bipedal footsteps behind you, don't take it for granted that it's your neighbor because it just might be a Dogman!

Vic Cundiff

Host, Dogman Encounters Radio

Contact@DogmanEncounters.com

INTRODUCTION

WHY WEREWOLVES AND DOGMAN CAN EXIST

Why Werewolves and Dogman Can Exist The majority of people don't believe in werewolves. However, after multiple years of research, I've come up with two simple concepts that prove one cannot say werewolves don't exist, at least not if they want to be taken seriously. It may not prove werewolves exist, but it makes any evidence to the contrary invalid. First, The Debunking Paradox:

"There are those who wish to try and prove that certain things (creatures, objects or concepts) are or are not possible. In the case of creatures and objects, one would be trying to prove that it does or does not exist. While, because of the Universe Example, trying to prove something impossible or nonexistent would be incredibly difficult, there are still those who may try to prove their beliefs.

However, there is a brick wall that cannot be passed while trying to do so. The paradox of trying to prove something impossible or nonexistent is that the only way you could know what to disprove is if the creature, object, or concept existed. Were this the case, the entire argument that it is impossible or nonexistent would become pointless. For example, in the case of werewolves, one could try to prove whether or not the full moon can cause any being to physically transform into another shape. The modern belief is that the full moon causes werewolves to transform. Therefore, this person makes the assumption that, if they can prove the moon cannot cause such a transformation, then werewolves cannot exist.

However, this is assuming that, if werewolves existed, the full moon is the actual cause of their ability to shift. The key phrase in the past sentence is "if werewolves existed". What if the full moon has nothing to do with the transformation? This is a possibility and, if there is more than one possibility, then we cannot say one is right over the other.

The only way we could know whether or not the full moon has anything to do with werewolves transforming is if they existed. If they existed, as said before, it would be pointless to try and state that they didn't. In the end, the only thing the person would prove is whether or not the full moon can cause a being to transform.

And even if they prove that it can, there's still no guarantee that it applies to werewolves. Every story before Hollywood came into the picture says nothing about the full moon causing the transformation. Therefore, again, we cannot know whether or not the full moon has any sway on a werewolf unless werewolves existed. In summary, it is pointless to try and prove something impossible or nonexistent because of this paradox." Second, the "Universe Example", which basically states that, because we cannot prove that something existing on this planet doesn't exist on others (yes, it sounds weird), we cannot say that the thing doesn't exist until we search every planet in the universe. Essentially, we can't say something is impossible or nonexistent until we've searched the entire universe and found no evidence of that thing's existence.

It's difficult to describe these concepts, but they are all that's necessary to debunk all claims that werewolves don't exist.

Ancient Wolf

By

Dogmeatsausage

Chapter One
A Historical review of the Black Dog
~Joedy Cook~

The current interest in creatures such as the Michigan Dogman or The Beast of Bray Road is nothing new. Reports of similar creatures don't simply date back years, they date back millennia. Even leaving aside the images of beast headed gods such as Anubis or Set in Egypt, there are accounts within the historical record of beings resembling dog-or wolf-headed humans. We can start by pointing out the strangely detailed account of a race of Dog headed men known as Cynoscephalae (Greek: "Dog-Heads") who lived in the mountains of India that is found in the existing fragments of Ctesias' Indica. Ctesias was a Greek physician of the 5th Century BC. In his work, he claims that this tribe dressed in animal skins, and cannot speak, but instead communicated by barking like dogs. He describes their teeth as being larger than those of dogs, and their nails were more claw-like, being longer and rounder than normal.

Ctesias describes these beings as being 'extremely just', and says that they can understand the Indian language even though they cannot speak it. When they must communicate with those who do not understand their 'barking', they use hand gestures.

He goes on to say that they are called Calystrii by the Indians, and that they live on raw meat. This description is typically dismissed as a fantasy, and many other features of his work seem equally fantastic. Some historians have argued that the Cynoscephalae are not literally 'dog headed' people, but instead refers to an especially low caste of people who 'live and eat like dogs'. The details he gives however, are oddly detailed if this was the case. He even estimates their population as number around 120,000.

However, he is not the only Greek writer to comment on these creatures. The explorer and ethnographer Megasthenes (from the 3rd Century BC) also mentions these creatures in his similarly titled book Indika. He repeats some of information that we find in Ctesias, that they have the heads of dogs, are armed with claws, dress in animal skins, cannot speak but only bark, and says that they have 'fierce grinning jaws'.) He also adds some further information that was lost from the existing copies of Ctesias (but cites him as the source of this information), that the females bear offspring only once, and that their children are white-haired from birth. The famed Greek historian Herodotus reports that ancient Libyan tribesmen also claimed there were 'dog-faced' creatures living in the wild to the west of them.

He does not name them as being human in shape, however it is an odd distinction to call something 'dog-faced' that is not actually identified as a dog.

During the 5th Century AD, no less a luminary than St. Augustine of Hippo speaks of the Cynoscephalae. He muses on them in his groundbreaking theological work The City of God. "What shall I say of the Cynocephali, whose dog-like head and barking proclaim them beasts rather than men?" he writes. In his conclusion he adds "Wherefore, to conclude this question cautiously and guardedly, either these things which have been told of some races have no existence at all; or if they do exist, they are not human races; or if they are human, they are descended from Adam." Augustine hedges his bets here, and refuses to define the Cynoscephalae as being either human or animal, a question that still exists today.

Indeed, the church seems to have something of a history with these creatures. One story of the warrior Saint Mercurios describes how the saint's grandfather had been killed and eaten by 2 dog headed men and how the saint then converted them to Christianity. He was said to use them as his 'special weapons' that he used against the enemies of Rome.

Like St. Augustine, church members often seemed to wrestle with whether or not these dog headed men were descendants of Adam or not. Early church father Ratramnus wrote to the presbyter Rembert about them in his Letter upon the Cynocephali, in which he decides that they are degenerated human descendants of Adam. This is in contrast, he claims, to church belief, which by this point held them to be beasts. Interestingly, he suggests that some may even qualify as being baptized through the unusual method of being rained on.

He tells Andrew that he will send to them "a man "of terrible appearance, whose face shall be like unto the face of a dog." Just as he prophesied, the two apostles found themselves outside 'The City of Cannibals' and an angel appeared to one of the men of the city, who had the head of a dog, and commanded him to go and protect the two saints from harm.

In a darkly humorous turn for early gospels, the dog-man asks if he will be supplied with enough men to eat if he does so. In this case, God gives the man the 'nature of the children of men', and restrains the man's bestial tendencies. Like other dog-headed men, he is described as being unable to speak human languages, but only make gestures with his hands.

Perhaps most interesting in this account is the description of the man, which matches in many ways the current descriptions of dog men or 'werewolves' sighted in the modern day. I'll quote it in full: "Now his appearance was exceedingly terrible. He was four cubits in height, and his face was like onto the face of a great dog, and his eyes were like onto lamps of fire which burnt brightly, and his teeth were like unto the tusks of a wild boar, or the teeth of a lion, and the nails of his hands were like unto curved reaping hooks, and the nails of his feet were like unto the claws of a lion, and the hair of his head came down over his arms like unto the mane of a lion, and his whole appearance was awful and terrifying."

This is the first account I have seen of the glowing yellow eyes that seem common in modern dog man sightings. The height given here is unusually tall, being close to 7 feet, while modern dog man sightings tend to be slightly smaller. The dog man that accompanied Andrew and Bartholomew admitted that his name was Uasum, meaning 'Abominable', until Andrew rechristened him 'Christian.'

When they got to the city of Bartos, the dog man covered his face so as not to frighten people. However, when the nobles of the city ordered that the apostles be thrown to wild beasts, Christian prayed to God to return his bestial nature, and when it returned, he slaughtered the beasts and ate them with such ferocity and savagery, that some 704 people are said to have died of fright.

However, when the Governor of the city begged for mercy and protection from Christian, the Apostles restored his human nature. Once restored, the dog-headed man felt regret for their deaths, and asked for those who had died of fright to be resurrected, which they were. The savagery of the dog-headed men remains as consistent as their descriptions, throughout history, as does their craving for raw meat or human flesh. Likewise they are always considered to be just as intelligent as any other people, simply incapable of communicating with them without the use of sign language, unless there is some supernatural intervention which enables it. The Old English Passion of Saint Christopher explains that prior to his conversion, Saint Christopher was 'of the race of the Dog-heads', which were said to 'have the heads of dogs and to eat human flesh.

Christopher could 'only speak the language of the Dog-heads' but was devout, and spent his time meditating on God. Finally he prayed for God to grant him the gift of speech, so that he might convert people to Christian worship, whereupon an angel appeared and struck him in the mouth, granting him the ability to talk as other men. There are several icons to St. Christopher in this form that are found among the Easter and Russian Orthodox churches. Some of his Russian depictions have him shown with something more resembling a horse's head, whereas Greek iconology tends to be more dog-like. Christopher has a great many stories about him, many of which deal with his appearance. In some stories, he was consider to have an almost unearthly beauty, and he prayed to God to make him ugly, so that people would be more swayed by his words than by his appearance.

In other stories, he was called a giant, who towered above other men. In the Anglo-Saxon Martyr ology, Christopher is said to have come from "the nation where men have the head of a dog and from the country where men devour each other." For his description, it is nearly identical to that of 'Abominable'; "He had the head of a dog, his locks were exceedingly thick, his eyes shone as brightly as the morning star, and his teeth were as sharp as a boar's tusk."

Dog-headed figures appear surprisingly often in early Christian art. They are often used to signify people from distant places, and we see examples where Dog-Headed people attend the Pentecost or listen to the teachings of Christ. It is clear there was a widespread belief that such - creatures existed, even aside from the tales of werewolves that began to appear in the Middle Ages.

Dog-headed peoples were often found in in medieval bestiaries and encyclopedias. Indeed, the book that had become known as The Newell Codex or The Beowulf Book contains many mentions of such creatures. In fact, there is some speculation that the most memorable enemy of Beowulf himself, the monster Grendel, may have had either lupine or canine qualities, based on certain descriptions and word choices in that poem. The Wonders of the East is a sort of travel guide for the near and Middle East which is also in The Beowulf Book. This book speaks of a place where 'half-dogs' dwell, called Conopenae. They are described as having "horses' manes and boar's tusks and dogs' heads and their breath is like a "fiery flame."

Also included in the book is a document called Alexander's Letter to Aristotle, which purports to be a letter from Alexander the Great to his tutor Aristotle, while on campaign in India. In it, Alexander claims his army was attacked by Cynoscephalae, however, they were driven off by Alexander's archers. The peculiarity of The Beowulf Book is that the first five of the books contained in it; The Life of St. Christopher, Wonders of the East, Alexander's Letter to Aristotle, and finally Beowulf all appear to have been originally bound together, while the remaining books appear to have been added later on. Historian Andy Orchard suggests that the original book was assemble thematically around monsters. I would go farther and suggest that the theme is actually Dog-headed men.

By the 13th Century, the popularity of the Cynoscephalae was starting to wane somewhat, and werewolves began making their appearance in romances across Europe. And yet, even the illustrious Marco Polo wrote of an island whose inhabitants had the heads of 'mastiffs' and who ate anyone they caught who was not of their race. Cynoscephalae made sporadic appearances after this, but as Europe moved into the Renaissance, the belief in them began to be dismissed. Based on the popularity of such romances as Bisclaravet and William and the Werewolf, the naturalistic Cynoscephalae began to give way to the more supernatural werewolf in the popular imagination.

However, the werewolves of the late middle ages and early modern period are quite different than the dog-headed men of prior tales. Apart from their supernatural aspect, the werewolf most often appeared as simply a very large wolf, rather than the figure of a man with a wolf or dog's head. 'Lycanthropy', or the belief that one could physically transform into a wolf, was even recognized as early as the 14th century as a form of mental illness. Even the judges and jurymen that presided over the trials of suspected werewolves during the era of the Great Witch Hunts across Europe recognized it as a form of insanity. However we hear no tales of dog-headed men during this time. Historians considering the Cynoscephalae tend to interpret them as being something akin to 'the ultimate foreigner'; something to represent persons coming from very far away. They also can represent particularly savage people…

in some case, barbarians, for example. There is certainly good reason to accept these interpretations in many cases. Dog-headed people are represented in medieval art, often positioned to represent those coming from the most distant places in the world. Others believe that these creatures are either mythical, or misinterpretations of descriptions of ordinary men that have crossed through several different language barriers before being recorded.

There is also reason enough to accept that this is often that case as well. However, it is intriguing to note that the description of them remains consistent across centuries and through many cultures. This creates a unique, unvaried picture that doesn't seem to exist with similar, mythical entities. The Cynoscephalae were savage warriors, carnivores if not outright cannibalistic, they stood upright, were capable of using weapons (indicating an opposable thumb), dressed in animal skins, spoke their own barking language,

but could understand human speech, had claws on their hands and feet, had tails, and lived in mountain caves, or more rarely, cities, in Northern India and North Africa. They had shaggy, mane-like hair around their head and shoulders, and eyes that seemed to glow. They found their food through hunting. Taken as a whole, this description matches the modern ones given by witnesses to the dog man or werewolf cryptid phenomenon. While it's impossible to know for certain if they are related, they seem at least physically identical to the tales of the Cynoscephalae. It is, at the very least, food for thought, and helps to develop a possible historical connection. More importantly, it gives us a consistent picture of them through time, which might be able to shed some light on the modern sightings and behavior.

The Jachal God Anubis

Helmet and collar representing a wolf, at the Museum of the Americas in Madrid. Made of wood, shell and made in the 18th century by Tlingit indigenous people, from the North American Pacific Northwest Coast. Tlingit people admired and feared wolves for their strength and ferocity.

Sculpted head of the "Loup Garou" or werewolf in Cognac, France.

Werewolf :

(Egil's Saga Exhibition in Borgarnes, Iceland)

Indo-European comparative mythology

Chapter Two

Werewolf of Germantown
By
~Adam E. Davis~

Germantown-Liberty Rd/Stacey Rd, Montgomery County Conditions- Clear night, bright full moonlight Driving back from the "Land of Illusion" haunted house attraction near Middletown, OH my friends and I decided to take a trip through where we had been hearing howls off and on when we would go out trying to capture audio of the howls that had been recorded near Liberty, OH. Until that point the recordings we had were barely audible, and the recorder primarily picked up night insects. Knowing that it had been chilly for a few nights, we decided to try our luck getting a recording, since most of the bugs were dormant now, and wouldn't be an issue on the recording.

 We turned north onto Germantown-Liberty Rd from Rt. 4 around 11:30 or so, and started traveling a good deal slower than the speed limit with our windows down about 1", so we could listen for anything. Germantown-Liberty Rd takes a "jog" of about half a mile east where it dead ends into Hemple, but a connecting road starts, called Stacey Rd. About halfway up Stacey, where it starts curving to the right, I caught a "glint" of eye shine from the right side of the road.

It was mid October, and I half expected to see a deer ready to cross the road in front of me since it was hard into deer season. Instead of seeing a deer standing by the side of the road as we turned toward the direction of the presumed deer along the gradual curve, next to a small tree on the shoulder of the road, we saw what initially struck us as the biggest dog any of us had ever seen. It was sitting/squatting in the ditch along the side of the road on its haunches, but instead of its "forelegs" supporting its upper body, it was more or less balanced, with its forearms resting on its thighs and sort of hanging between its legs. If I had to reference anything for how it was squatting, it was similar to how a baseball catcher squats behind home plate.

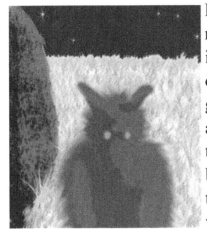
I slowed the van down even more as we looked at this thing in disbelief, with its head more or less centered in the passenger side window, staring back almost like it was looking through us. Its hair was longer but not shaggy, somewhat matted, and its shoulders were very wide as we looked at it dead on from the front. A dog's shoulders aren't a whole lot wider than its chest, this thing's shoulders were easily twice as wide as its rib cage, so it was more proportioned like a line backer. In the moonlight, though it was a hairy creature, you could see it was very heavily muscled.

It didn't seem to even twitch as we crossed in front of it until the back end of the van just cleared where it was sitting, then it stood up, taking two easy steps across the road behind us. I watched it in the rear view mirror as its shadow passed across the back of the van, the others turning to look at it as scared and awe struck as I was, and I saw it was easily head and shoulders taller than my van(1998 Plymouth Grand Caravan for the record). That would put its height at well over 7' tall, to 7'6" in our estimate. It took those two steps from the ditch to the other side of the road in just under a second and a half, and was gone.

 We drove on and turned back north onto Germantown-Liberty Rd where Stacey dumps back into it, and were hoping to catch sight of it again, or possibly capture a howl or anything from it, but at that point it was gone. We attempted to find prints on either side of the road, but only found where the grass had been mashed down, since the weather had been dry for over a week prior. We attempted to emulate how it crossed the road with such ease in two steps, and even standing on the edge of the road, we couldn't cross the road in less than 3 steps taking awkward giant steps, let alone 2 steps from the ditch it was squatting in, a good foot off the side of the road. No other tracks were visible on the opposite side of the road in the stubble of whatever was growing there before.

My Second sighting of the creature was on April 2006

4617 Diamond Mill Rd, Montgomery County, OH .

We were in the house upstairs with a window and the door out to the balcony open to let some night air in. While watching TV, we heard a howl from outside. We had attempted to record other similar howls before with no luck since they would stop before I could grab my recorder and get outside.

I managed to get my digital recorder and stepped out to record them this time, though they had already been howling for a good 30 seconds or more. I was able to get several howls recorded before the local dogs started barking across the road and down the road from us. The closest dogs were about 175 yards away at my 5-6 o'clock). This went on for over 2 minutes before the howls died off, which were coming from 3 directions at once, anywhere from about 400 yards from my 5 o'clock. I was facing due west toward the open field behind the house) to roughly 1200 yards at my 10-11 o'clock, the other one(3rd source of the howls) was somewhere between that range, at my 2 o'clock.

These ranges are estimates based on how loud they were in relation to my elevated position about 14 feet up on the balcony, and sound traveling over more or less open ground.

The following are Newspaper clippings from the Dayton Daily news :

Howl in the night stumps Liberty

By William G. Schmidt
For the Dayton Daily News

JEFFERSON TWP., Montgomery County — It is a sound that's likely to cause the hair on the back of your neck to stand up.

It's an eerie howl that has been waking some residents in a rural area just south of Liberty this summer. It's turned a few heads of residents coming home late at night.

The question in everyone's mind: what is it?

The source of the ghostly wail, at least, is pretty well pinpointed to between Liberty-Ellerton Road and Germantown-Liberty Road — a very natural area with a stream flowing from it into Bear Creek. It's a perfect wildlife habitat.

Rob Seiter, 55, who lives on the

ONLINE EXTRA

Listen to the howl and tell us what you think is making the noise at DaytonDailyNews.com

west side of the area, describes being awakened between 3 and 4 a.m. He and his wife, Mary, also have heard the sound earlier at about 11 p.m. — but always at night.

"We thought it was a bird," he said. "Then it sounded like something ripping something apart."

Bravely, Seiter decided to see if he could uncover the cause.

"I went out one night and there were two of whatever it is. I never saw them but they were definitely

PLEASE SEE **HOWL** ON **A8**

Newspaper clipping from the Dayton Daily News
By William G. Schmidt September 2005

Junichiro Koizumi prays at the Chidorigafuchi National Cemetery where the remains of unknown victims of World War II are honored. from a shrine criticized for glorifying the aggression that provoked two atomic bombings.

The war's legacy lingers in Asia, where many of Japan's neighbors

the late 1890s.

In Seoul, South Korea, a joi delegation from the Koreas a nounced Tokyo after visiting prison where Korean indepe dence fighters were tortured a

HOWL

Shakes residents of rural area

CONTINUED FROM A1

on the ground."

Being close to the source, Seiter said he heard the "call" end with a gurgle, even a 'baby growl.'"

Is it frightening?

"Oh, yeah!" Seiter said. Mary Seiter even obtained a copy of a tape, from a neighbor, with the shrieking sounds.

The Montgomery County sheriff's office reported no calls from residents frightened about the sound.

Doug Horvath, a naturalist at Germantown MetroPark Nature Center said he is stumped. It doesn't sound like any bird he's heard. He said there have been bobcat sightings in the area but it doesn't sound like that either.

Betty Ross, the director of the Glen Helen Raptor Center in Yellow Springs said, "Sounds human-like to me. The next closest would be coyotes, but even that doesn't

fit, except for someone imitating them. It definitely doesn't sound like any birds I know, especially not raptors, and they are the ones I am most familiar with."

A neighbor to the east, Bonnie Maschino, 53, has heard it twice and thought it sounded like an injured animal. Still, although she's seen deer, raccoon, even a blue heron in the area, none seem to fit the bill.

Her next-door neighbor Bill Schlater, 54, says he thinks he knows what it is: a red fox.

"I got on the Internet and found a red fox warning sound. It was real faint but it's got the same sound," he said.

Schlater said he's seen young foxes in the area and even filmed them playing in his yard. He raises chickens for meat and that, he said, may explain some of the attraction to the area for the foxes.

While Schlater hasn't heard the wails since June 19, Seiter said he was still hearing them in early August, though fainter.

Schlater said even when the sounds were going full-throttle, he could go outside and clap his hands and get them to stop.

He took it all in stride as simply a hazard of living in the country. "We get lots of noises."

RELATIVES LEAVE an Athens mo airliner flying from Larnaca to Athe

Police in

Coroner says some alive when plane crashed

Ohio districts, schools improve

Designations	Districts			Schools		
	02-03	03-04	04-05	02-03	03-04	04-05
Excellent	85	117	111	630	920	889
Effective	177	229	297	771	906	1,136
Continuous Improvement	278	224	175	1,242	1,211	962
Academic Watch	52	34	21	237	125	239
Academic Emergency	16	4	5	338	222	288

Source: Ohio Dept. of Education

Continued from A1

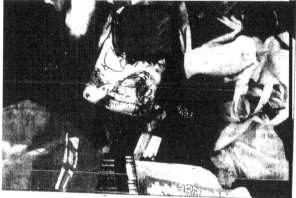

kbags that will be given to children from DMHA sites at a cookout today.
ED ROBERTS/DAYTON DAILY NEWS

e away 1,001 bookbags

nong the hous-
r would never
s," dividing the
kewise, DeSoto
cross," the 54-
resident said.
and others in
er Ministries
r not only fam-
wo housing ar-
n, but four oth-
olitan Housing
well for an af-
ip, food and a
ags filled with

BACK TO SCHOOL

school supplies.
The back-to-school event will begin at noon on land at Gettysburg Avenue and James H. McGee Boulevard owned by the church.
"Last year we blessed over 1,000 kids, with no incidents," Mitchell said.
Revival Center Ministries' Winning Souls for Christ Outreach

Ministry has spent months visiting families at Dunbar Manor, DeSoto Bass, Hilltop Homes, Arlington Courts, Mount Crest Court and Parkside Homes.
Mitchell, a minister, works alongside her husband, Elder ... daugh... ole Freeman, 34. O...seeing the ministry is Senior Pastor Paul Mitchell and Keisha Mitchell. The church is at 3011 Oakridge Drive.

PLEASE SEE **BOOKBAGS** ON B4

JEFFERSON TWP., Montgomery County — Those who have heard the Jefferson Twp. howl are certain: Whatever is making that noise is not human.
But after John Dinon, director of animal conservation programs for the Cincinnati Zoo, listened to a tape of those howls for the *Dayton Daily News*, he is equally sure: It most certainly is human.
"I can't identify it as any animal that I have ever heard in my life," Dinon said. "I believe it is a human being, because that's what it sounds like to me."
Only two animals found in this area howl: dogs and coyotes. And the tape is neither, said Dinon, who has been in the zoo business for 20 years, holds a bachelor's degree in animal science from Michigan State University and has been with the Cincinnati Zoo for 3½ years.
Dinon said a colleague believes the tape sounds like a download from a Bigfoot Web site.
"It could be that Bigfoot is in Dayton," Dinon said. "I'm skeptical."
The debate about the sounds recorded on the tape, which is on www.daytondailynews.com, continues to rage. Animal experts and Miami Valley residents are divided about what could be making the noise.
The tape was recorded by a Jefferson Twp. resident and given to the *Dayton Daily News* by Rob and Mary Seiter, neighbors of the resident. They also reported hearing the howls many times during the summer, and said the tape represents what they heard.
Mary Seiter said she has no doubt that the tape is real, made from her neighbor's hand-held recorder, and not downloaded from the Internet. Those howls went on for weeks, Mary Seiter said Friday.
"There's no human with that much dedication," she said.
She called the sounds "unreal" and said many neighbors had heard them. "There's too many of us," she said. "We're not nuts."
Lee Coffey, a Sidney resident, feels their pain — and said he

PLEASE SEE **HOWL** ON B4

med
ector

DC's first of-
nder her ten-
act for Oprah
Angelou at the
g Arts Center
0 tickets re-

58, can't take
t the former
and WRNB-
on hopes to
mpany's pro-
ty where she
hed a career

Second Newspaper clippings fro Dayton Daily news October 2005

HOWL

Wolves can raise a ruckus

CONTINUED FROM E1

ter Burnett, the quintessential plugged-in 1950s Chicago bluesman. The 1991 Chess box set would be the best to get from this late great.

▶ **Wolfman Jack**: Where were you in '62? That was the year a young radio announcer from Brooklyn named Bob Smith started becoming famous for his trademark howl, gruff growl of a voice and jokey moniker, Wolfman Jack. He was heard far and wide on the super-powered Mexican-border station XERF back in the wild and wooly days of early rock radio. The most famous DJ ever? Well, one of the few we can think of who had a song written for him by the Guess Who in 1973. Jack died in 1995.

▶ **Bob Dylan**: All Along the Watchtower remains one of our favorite tunes by a man who has done his share of growling and howling over the decades. Here's how it ends: "Outside in the distance a wildcat did growl. Two riders were approaching, the wind began to howl."

▶ **Oscar the Grouch?**: Perhaps watching the videos Sing-Street Animals or Bugs Bunny's Hoot & Howl With the Sesame Street Animals or Bugs Bunny's Hoot-O'ween Special would provide some clues to the identity of the howler.

▶ **The Hound of the Baskervilles**: The fearsome dog on the Scottish moors — "a foul thing, a great black beast, shaped like a hound, yet larger than any hound that ever mortal eye has rested upon [with] blazing eyes and dripping jaws" — starred in the Sherlock Holmes novel from 1902 by Sir Arthur Conan Doyle.

▶ **Real, live wolves**: Yes, of course they howl, but there aren't any around here. Thanks to federal protection and repopulative efforts, though, wolves are making a comeback in parts of the United States. The International Wolf Center (www.wolf.org) estimates about 4,000 gray wolves in the wild, mostly out west and in

the far-upper Great Lakes region. Fewer than 100 red wolves are believed to live in the wild in this country, mostly in the Southeast.

▶ **Rowdy Raider**: Wright State University's prowling mascot lets loose a periodic, way-spooky aaaahhhh-ooooo as the basketball team takes the floor for an opening tip at the E.J. Nutter Center and Wildlife Preserve.

▶ **Thurston Howell III**: Superbly portrayed by the late Jim Backus (he was the voice of Mr. Magoo, too) on Gilligan's Island. The uncharted island's resident millionaire and his lovely wife Lovey (Natalie Schafer) — were always putting on the dog in their endlessly stylish wardrobe that somehow survived the Minnow's demise.

▶ **Our best guess**: The howl, mournful response to prices rising yet again at the gas pump.

Contact Ron Rollins at 225-2168.

▶ **Howl**: English majors will recall the famous and famously psychedelic poem by Allen Ginsberg that sprung from the Beat Generation and was considered obscene when it came out in 1956. Ginsberg's opening lines: "I saw the best minds of my generation destroyed by madness, starving hysterical naked, dragging themselves through the negro streets at dawn looking for an angry fix, angelheaded hipsters burning for the ancient heavenly connection to the starry dynamo in the machinery of night..." Hey, we didn't say it was The Raven, OK?

▶ **Howlin' Wolf**: Aka Chester Burnett, the quintessential plugged-in 1950s Chicago bluesman.

Artist Rendition of the Germantown Werewolf©
NADP
By
Sybilla C. Irwin

Chapter Three

BEAST IN THE FIELDS
~Anthony J. Chaney~

It's February 2010 and I had just left Florida, having now entered that little bottom portion of Alabama on I-10, that long stretch of highway which one could take from the East Coast and travel all the way to the West Coast. That was my goal at the time. To drive from Fort Myers, Florida and all the way to Olive Hurst, California. It is said that most will end up once in their life time making a long journey. Life is that one time journey. But there are some journeys we take in life that can leave us having a story to tell. Mine is such a one.

So I am traveling on Interstate 10 and I am getting really tired. With little money to work with, I had decided it best to use rest stops, gas stations, and truck stops to pull over in so I would be able to get some rest and sleep. It just happened that my first place of rest would leave me with an incredible story but, one that I would forget. That is until I remembered it most recently. But only because it relates to my story, my sighting of a upright walking canine cryptid. I understand why I had forgotten it.

It was so unbelievable at the time. Now I don't easily dismiss others when they tell me about strange creatures they have seen.

So I pulled into this rest stop some where in Alabama. Some where past Mobile yet near the Mississippi border as I recall it. Pulling over to the rest stop, it was quite and there were no other vehicles around. I was the only one there. It was night by the time I had parked my Chevy Astro van. I walked around the area so I could stretch my legs. As I did, I noticed the thick woods lining the back area of the rest stop. The rest stop area was neatly kept. Grass mowed and bushes trimmed. But the back area was just thick and dense forested area.

I noticed this maintenance access road. There was a gate and it was open. I went to walk over so that I could explore the area. But then a patrol car pulled into the parking lot. It was a security patrol service. Having a background in security work, I approached the officer and started making conversation with him.

He appeared to be kind of nervous. He would always look off towards the wooded areas. Then he told me to be careful at these rest stops at night. That they have had problems. There was some deaths as he related the following story to me. He told me that he had other partners that were posted up at other rest stops along interstate ten because that there have been some gruesome deaths.

As I stood there listening to him, I couldn't but help notice how he would stay within the light that shone down from the available lamps which sat in front of the public restroom building. At this time, I caught him telling me to stay in well lit areas at the rest stops.

They don't like the light he said. Who is they? I had asked him being curious. I was so tired and yet found it hard to pay attention. Until he said something that threw me off. There are werewolves that live out in the woods. But the local authorities just blame it on a serial killer. He told me how that one of his partners had seen one before. I told him that I was sleeping in my van tonight. He told me lock my doors and stay inside my vehicle until I leave. He was also glad that I was parked right in front of the lights. Being to tired and not sure about how to handle what he had told me. I told him good night and that I was going to bed. But not before he asked me to watch him first. He had to close the gate to the maintenance access road that was flush with the woods.

I watched him get into his patrol car and drive up to the gate. He could of walked over there but chose to drive up to the gate instead. After closing the gate and locking it, he drove back up to the parking lot where I stood waiting and watching. He thanked -

me and then I went into my van and locked the doors. Exhausted, I fell asleep pretty fast. I woke up early in the morning. The sun was just rising and there was enough light to see the area. Especially with a fresh pair of new eyes. The security officer was gone. I had wanted to ask him about this whole werewolf thing after I had rested. But to late. I missed my chance.

It's February of 2011. Sometime in February of 2010, I had had made it into California. There I went into security training and had gotten myself work. I was glad that I was doing private protection work again.

So after spending several months of doing weekend work, I had found myself protecting a dance club up in Chico, California. There, five partners and myself would remove any trouble makers and gangs when they caused problems. We broke up plenty of fights. I lived in Marysville, California at the time so I didn't car pool with my partners who lived in Yuba City. They would travel through Live Oak and Gridley to get to Chico. I would drive through Oroville. I enjoyed driving through the area because I would get to see the forests. I love the outdoors and I would do a lot of fishing at night. But that all had changed for me.

After a long Saturday night at this dance club, I slowly climbed into my car. I drove slowly out of the parking lot and then through the city, making my way towards the highway. Once on the highway, I made my way towards Oroville.

I was driving slow because I was tired. That and it was Sunday morning. Sometime around 3:00 or 3:30 in the morning as I can recall it to the best of my ability, I came into Oroville. I was about a mile or less before Ophir Road as I drove my Chevy Malibu towards Marysville on state highway 70.

Up ahead and just behind or next to a road sign, I had seen some movement. I kind of slowed down to get a better look. As I got closer, I observed a nine to ten foot tall hairy person. I rubbed my eyes and gave another look. I slowed down some more again. There were no other vehicles on the highway at the time because it was a early morning Sunday. I sat up more in my car and leaned towards the windshield as I looked at this person. Then I got kind of worried. I observed this nine to ten foot tall werewolf looking right at me. I could not believe what I was looking at. But there it was. I slowed down even more, just so that I could make sure it wasn't some kind of cardboard stand up.

About sixty but maybe eighty feet away. It moved! But casually. It just turned and took several steps out into this field behind it. Then dropped down to all fours. It was massive in size. It had brown fur with light brown or tan streaks in it's fur.

Some what reddish but probably a rich brown. It's head was big. Almost like that of a bear but with a muzzle being around ten to twelve inches in length.

It's legs were like that of a dogs. Yet its upper body and arms were like that of a human. It had big muscles and I could see the definition to the muscles as it ran through the field really fast. The fur was some what long because it was cold outside. I didn't notice any tail. I was to busy looking at the size of its head, shoulders and how it moved on all fours. I could not believe the height I was seeing when it was on all fours. I would say it was around four to five in height from the ground when it was on all fours. It was really long looking as well. Little longer than the length of my car I believe.

About half way through the field, it had stopped briefly to look back at me. At this time, I was standing on the seat of my car with my arms rested on top of the car. I was outside watching it because I could not believe what I was seeing. When it had turned to look at me, I climbed back into my car really quick. I thought at first that it was going to charge me. But it didn't. I put the gear into drive but kept my foot on the brake peddle. I continued watching it as it ran towards a tree line which borders the back of the field and the Feather River which runs into the Oroville Wildlife Management Area. Taking my foot off the brake, I took off out of there.

. I was wide awake and the whole time I finished my drive home to Marysville, I had the thoughts of what I had just observed running through my mind. I just seen a werewolf! A beast in the fields. I could not believe what I had just seen. My mind had a hard time processing it. My mind was trying to rationalize the creature. It was a horse, no, it was a bear, nope, you actually had seen a werewolf so get over yourself. I will never forget !

Sometime in 2012, I was still talking about it, especially with my wife. I had done a lot of research and came across many researchers names. Many accounts of people seeing what has become termed Dogmen. It just so happened that during this time I was discussing these things with my wife, that one of my oldest sons who was visiting with me, overheard my conversation I was having with his step mother.

He broke in and told me that he had something to tell me. I never imagined that it would be related to what I share now. Sometime after my sighting of this creature in 2011, my son who had lived on Aspen Way in Olive Hurst. He had gone into the garage so he could do some laundry. He noticed his dogs huddled together who were laying on the couch. Shaking and silent, they stared at the side door to the garage. Curious. My son told me that he had looked over to the side door which was only ten feet away from him.

That is when he observed a black furred werewolf step easily over his six foot fence. Entering the side yard, it passed the side door that him and his dogs were watching.

Going straight into the back yard, he lost sight of it. My son told me that it was all black. That it smelled like a very bad skunk. It had dog shaped legs. It was very strong looking because it had a lot of muscles. He could see the bottom of it's muzzle because it's head was just above the door way. We estimated it to be around seven in a half or maybe eight feet tall. He never said anything because he thought people would think him crazy. I know the feeling because of my experience with this after I told some about my sighting of a werewolf back in 2011.

When my son had told me this at that time. I was concerned that the one I had seen could have been a alpha male who sent a beta male after my scent. But this wouldn't be the case I would learn later from my son. Apparently, there was strange things happening in the field behind his house. Just behind his backyard is a field. A few feet away from his back yard fence is a six foot deep ditch. One night while camping in his backyard, he observed glowing amber colored eyes inside the six foot ditch. They were a foot above the ditch! Meaning that something that was around seven feet tall was watching him. He had told me that this occurred before my sighting. Sometime in 2010.

Interestingly enough and just recently. A family came forward and recounted their story to me. They have given me full permission to tell their story but only under anonymity. They had seen the sketch of the werewolf or Dogman I had seen and placed into my book (KOHUNEJE: Sketches of An Encounter With Dogman. Pub. June 2015).

I call them the Smith family and this is their story. In 2009 while they had lived on Fifth Avenue in Olive Hurst, California which is about three blocks away from where my son had his strange happenings with a Dogman some time in 2010 and then 2011. The husband who holds a good steady job and works locally with a well known retail establishment related this to me. It was night time as he observed a eight in a half to nine foot tall werewolf like creature jumping from roof top to roof top on a couple of houses right across the street from him. It had German Shepard like head and dog legs yet, it had mostly human looking body. He observed long claws and that it had dark colored fur. He couldn't tell what color because it was dark outside. He said it was pretty buff looking and then he lost sight of it when it went right into a tree.

His wife told me that one night, she was sitting in the bedroom where she was facing a mirror while she was applying make up. That is when she seen a werewolf looking through her window at the back of her head. She turned to look at the window and seen that the creature was still there looking at her. She screamed for her husband and the creature took off. The Smith family moved soon after the incident.

The family revealed to me that they never said anything because their family, friends, and co workers would think that they were crazy. Unfortunately, there is a lot of truth to this reality. They were more than happy to share their story with me because they thought they were the only ones that had seen this Dogman creature. Being only glad that they weren't the only ones. With plenty of wooded areas and open spaces in the rural area of Yuba County, coupled with a vast amount of open fields. These creatures could easily move in and out of our rural area at night, in the dark, these beasts in the fields.

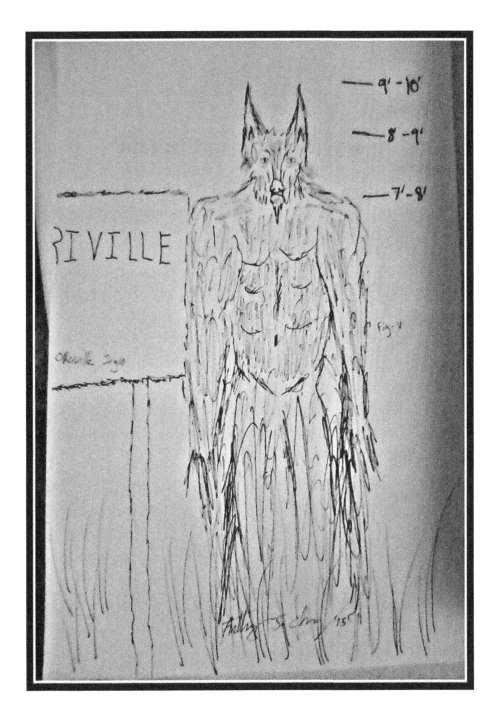

Eyewitness Drawing

Chapter Four

Bigfoot, Dogman and Human
~David Leidy~

It was back in 2009 in October. I remember the date very well, as it was my 49th birthday...10/10/2009. My wife, Lisa, and I are very fond of camping. Ever since we met, we have camped just about everywhere east of the Mississippi that has a National Forest and nice State parks. We decided that since tent camping was a game for younger bones, we entered into the realm of travel trailers. After looking at many, we decided upon an Airstream. They are really nice and expensive, but we intended to get a lot of use out of it.

We got a 25 footer so that we could fit into just about any campground. It has full hookups, but we really never used anything but the electric, and hardly that. Plus, we installed solar panels, because we intended to dry camp 100% of the time. We traded in my F250 and bought an F350 for the power. It's an old joke among Airstream owners: a $60K trailer pulled by a $50K truck, looking for a free campsite! Well, as I said, we were older now and had more resources and decided to go that way.

We embarked on a tour of the East Coast national and state parks. We sometimes would stay for a week or two or sometimes spend a day or two moving about; sometimes never unhooking. We both have a lot of outdoor experience. I was a boy scout and also spent a great deal of time hiding and camping, either with friends or alone. I was aware of strange things in the wilds, especially at night, but not sometimes in daylight. I found weird places for sure: Daniel Webster Forest in Vermont, the Pine Barrens of NJ, the Great Smoky Mountains to name three.

With my wife, we really experienced nothing really weird, except maybe with everyday people, as it goes. Well, one particular time, we decided to go to the Outer Banks of NC. We had no plan, per se, and decided to wing it. We traveled slowly, staying at campgrounds along the way, though always state parks, as they seem to be less commercialized and attract less "noise."

We made our way south and steered toward the coast. After crossing into NC from VA, we saw that it was approaching dusk and pointed out the closest park - Pettigrew State Park near the Pocosin Lakes NWR on Phelps Lake. As it got dark, we passed through a few very strange dark towns. A few miles before the park, we passed through Creswell. This was another dark town…really dark. I stopped and uncovered my off-road lights and continued.

Around a corner was a big house and as we approached, we saw about 15-20 people in the yard…in the dark. Before our lights illuminated the scene, it must have been pitch black there. We wondered why people would hang out in the dark like that. They all stopped dead and stared at us as we passed. The last guy said, "Good luck man." Uh oh, I said to no one in particular.

So we wind our way down to Pettigrew SP, just as the park ranger is leaving. There was only one other camper there and they were talking. The place was rather muddy and wet, after the rain and there weren't many trees, except by the lake; just swamp-type scenery. We asked how it was there and the ranger looked at the other guy and, (after looking around), said, "OK…pretty good. I promise to be back at sunrise." He seemed nervous and kept looking around. The other guy too. We decided to skip it. The other guy said, "You're not staying? Well, then neither am I. I'm not doing another night like last night." Ok…weird.

We left and drove toward Roanoke Island, as we figured we would just stay in a store parking lot, as we often did. There were no other parks nearby and we were not going to boondock in the swamp! So we headed down route 94, crossed the Alligator River, and drove into the middle of Alligator National Park. OK, having lived in PA my whole life and camping mostly in the North, large man-eating lizards were kind of new to me and I felt uneasy at the reality of a predator without any real "personality." To me, they seemed like eating machines with a real disregard to what the prey might be. A reptilian view? Not my thing! Ok, well while we drove through the night and we looked up the park on our phones, (amazed that we had signal). We read about the Red Wolf, a critically endangered species that was very rare. It was re-introduced into this park a few years back, but no one had seen one in a while. We were almost to Manns Harbor, near the ocean/bay, when we saw a dog calmly walking down the road toward us.

He never really looked at us, though I could see a yellowish glare to his eyes. We looked at each other and really couldn't believe it. It really looked like a Red Wolf! We just read about that! OK….

It was about 1AM, when I heard a loud "thunk" on the trailer. UGH, not here! I didn't really want to get out here, but decided that I should look anyway. I left the truck running, grabbed my 357, holstered it, and walked out. I checked the truck and the trailer and didn't see anything obviously wrong. Just a branch maybe? Then I saw a dent in the trailer, just below the right taillight. I checked it and while I was looking at it, the hair on the back of my neck stood straight out and all the insect noises stopped. I had been in the wilds enough to know what that meant. Someone or something was behind or near me, staring. Something odd. It was sort of like a predator/mountain lion/coyote feeling. I slowly turned around and didn't see anything at first. The feeling got stronger and I pulled the gun out.

I still saw nothing, but I began to hear rustling and water noises. I focused on the direction of the sound and I saw a dark shape behind some scrub bushes. As my sight sharpened I saw a somewhat familiar shape. It seemed like I was looking at one of my German Shepherds. Although, this one was reddish-black, very large, with tufts of hair above its ears, (which were both facing me). It had its front paws on a rock and it was crouching behind it. One of the ears was light gray and bent over somewhat.

Our eyes met and I got a chill. That was the look of a hunter with no emotion. It looked to Its left/my right and then moved slowly forward. Then it did something that haunts me to this day. I saw a deer do this once and I thought I was crazy. It lurched backward onto its rear legs and there were two distinct popping sounds, as it straightened up. Then I really saw him. He was somewhat muscular above the waist with hair like a mane around his shoulders. His hair/fur was very reddish, though dark with black streaks. He had yellow predator eyes, pointed ears, and hand-like paws with fingers and claws on the end of very long muscular arms. His lower half was thinner with strange canine legs and I thought that I saw a tail. Wow. This was the stuff of nightmares. Was I really seeing this? It made noise…a low growling and then it looked to its left again and I followed it gaze and I saw another one. It was really dark, on all fours and moving to my right…a flanking maneuver. Uh-oh… time to go.

The one in front of me took another step and showed his teeth. I could see them in the moonlight glinting, in fact, I can still see them right now in my mind. I backed away toward my truck and I glanced to my right. I still heard the other one, but I couldn't see him anymore. Not good. I reached the truck and I saw that the first one had reached the road and had one paw with claws on it. I took ail and fired at him. I am a good shot and I know that I hit his shoulder, as he turned at the last minute. I was aiming at his head because I felt in my heart that a warning shot would do nothing. It amazed me that he did not react to the shot as a human might. He took it like a punch and I had the distinct thought that I just made him angry! He did stop short though, as did the other one, and I used that time to get in the truck.

Honestly, I do not remember getting in! My wife was in a state of panic because of the shot, also because she had never seen me frightened before, and I was shaking. I do not remember putting the truck in gear and taking off. I was moving pretty well, as the truck was very powerful and always made light of the 8000Lb trailer. I looked in my mirrors and saw one on each side of the road and at least another one behind each of them! Wow...4 of them. They stayed with me up to about 30-35mph and then they slowed and just veered back into the swamp and were gone.

I didn't stop or really talk, until I crossed the causeway/ Bridge and was on Roanoke. I pulled into the Manteo Shopping Center and stopped. Then I looked at Lisa and I told her what I saw. She said that she didn't see anything, but admitted that she was looking at her phone when I was behind the trailer. I am so glad she didn't get out or see them. I do not think that I would have ever gotten her out of the house again. I decided not to go into too much detail because of that and I told her that it was a warning shot for a predator. (One that I told her that I had misidentified.)

Roanoke Island has its own history for sure. We stayed in a Grocery store parking lot that night. Many businesses allow overnight stays by trucks and campers. They figure that you will wake up and shop in the AM and that is usually what happens. In fact, Walmart and Cracker Barrel have special parking for trailers and they ask that you park so that the store cameras point toward your doors. OK, so we had an uneventful night…except that I had some really bad dreams of you-know-what. I kept seeing them really clearly and there were many more in the dream The thing that kept repeating in the dream and still bothers me to this day when I think of them, is the actual point in which they stand up. It is the popping sound. Like it's an abnormal move and they do it anyway. That and the fact that their eyes never leave you when they do it.

They seem to run faster and quieter on 4 legs, but seem to shift effortlessly from all 4 to a half crouch. They seem to have to stop completely to perform the erect stance and it takes about 15 or so seconds to "settle in." At least that what it looked like and my dreams were quite vivid also. So I woke, (in a sweat), and we went into the grocery store. Of course everyone was staring at us. But…that is not unusual. The truck, the Airstream, the circumstance, plus I am in my 50s, 6'4" tall, 260 lb., and my hair is past my waist, though usually braided. We get a lot of looks! Its OK, it never bothers us. Its just all part of the adventure….

Anyway, after we left, we put the groceries in the trailer. Lisa got in her side and I was walking around the trailer to my side. I saw a truck in the parking lot and the driver had me in a dead stare and it was decidedly unfriendly. There were several alternatives I had here and I chose a query. As soon as I began to speak, the guy completely fell apart and started apologizing to me. He said that I was a carbon copy of a friend of his and that he thought that he was messing with that guy. We both laughed about it and I asked him about the island and the surrounding area. He told me the normal tourist stuff and told me about a couple of good restaurants.

. I gauged him, looked at the rifle rack in his cab, and then decided to ask about what I saw in the swamp. He rocked back on his heels as if I struck him. He looked around and said, "Holy crap, I thought I was the only one!" We then proceeded to sit at the table in the camper and talk about it. It seems that he had exactly the same experience, twice. His name was Erik. He described them down to the bent gray ear on one of them. Damn. Same guy.

Erik told me that the 1st time he only saw one, but the second time, he saw 2, maybe 3 and that second time, he shot one. He said he hit in the chest and it didn't stop at all. Then he hit it again in the leg and it stopped and looked down. Then they all retreated a little bit. This had to be good to know, I guess. Erik said that they had all displayed hunting behavior. He said that he saw them on the mainland while he was hunting for game. He also said that he doesn't go there anymore. In fact, he said that he used to hunt everywhere in and around the swamp and it seemed to him that this "Dogman" thing has only come up within the last 5-7 years.

I thought the same thing as I had been all over, seen many predators and other things, but I had never them, although I never really looked for them. Erik agreed with that. Coyotes, Bobcat, Bear, Mountain Lion, Gator for him, and snakes were all predators we both knew. These "Dogmen" were new and different. He told me that a friend of his is a Bigfoot researcher and that he had been out with him. They actually had a good time doing it and that neither he nor his friend ever felt seriously in danger. The "Dogman" thing has really put a damper on things. It seems that wherever they are, the BF phenomenon seems to diminish.

Well, I bought him lunch and the 3 of us enjoyed it and we never mentioned it again. Lisa and I went to Fort Raleigh at the North end, then down to Kill Devil Hills, (I wonder now where that name came from?), and checked out the Kitty Hawk First flight area. I began to put the whole thing out of my mind. Being in civilization is real good for that. We just kept the trailer hooked up and we were like hermit crabs for a couple days!

We stayed one more night at Manteo and then we drove down to the outer banks. What a great place it is. If you haven't been there, you should go.

We went between seasons and it was quiet. We finally settled at the Frisco Woods Campground. It was closed, but I called the owner and he told me to stay as long as I wanted to and that the electric was on. Very cool…I was sure to leave a nice envelope upon leaving. We unhooked the trailer and set up our camp. It was great as we could see the back bay and the ocean at the same time. We were very much alone and had nice days and nights. I was wary of alligators and I had those dogmen in the back of my mind, but there didn't seem to be any predators on the OBS.

We stayed for a week and then headed back. Unless you take a ferry, there is only one way on and off. Right before we left the area, we dry camped, hooked, one more time. We were off of Roanoke and onto the mainland, (which I was wary of now). We stayed at a nameless business in Manns Harbor on Old Ferry Dock Road, (with permission of course). We didn't venture out after dark and slept soundly. Before we went in, the owner told us that we should stay inside after midnight though and said nothing more. He didn't have to tell me. Manns harbor wasn't far from the encounter place.

In the middle of the night, I got up for a pee break and decided to look outside. It is never a good idea…better to leave the blinds down! Anyway, I thought that I heard stealthy movement outside.

There was nothing…except that I thought that I saw two eyes on the edge of the woods and the insects stopped again. They were eyes because they blinked, twice. They were about 4-6 feet up, I'd say and after a while they were extinguished and gone and the noise started again. If it had been any other time, I would have dismissed it as totally normal for the woods. Even woods on the edge of a town. Now, I am nervous about the woods.

A couple hours later, toward dawn, it was silent outside, except for distant noises and I woke up again. I saw a dark shape move out of the woods and it wandered over to look at a rabbit hutch. I was wondering before why the hutch was completely enclosed in heavy fencing. Well, now I knew.

As I watched, the shape became what I thought to be a Dogman in a crouch. I couldn't tell for sure, but it was big and seemed to have a snout and a tail. It was panting like a dog. It tried to open the fence, but what struck me is that it did not tear at the fence…it stood up on 2 legs and tried to figure out the lock. It kept looking around, but it was so dark, I couldn't really see it well. Once, right before it dropped back down and went back into the woods, it looked right at me through my window. I swear that it saw me, because I got that icepick feeling in my stomach.

In the morning, we woke up, thanked the man and set about hitting the road. There was no sign of the creature or creatures anywhere. The guy saw me looking around and he looked at me knowingly, but said nothing. When I started to say something, put his finger against his nose. I knew that to be the British equivalent of putting your finger to your lips. I nodded and we got in the truck and drove away.

I wished that there was another way to the highway, but all roads seemed to lead deeper into the back country. As it was, I drove on Rt 64 toward Rt 32 back to the Great Dismal Swamp, Suffolk, and eventually to Norfolk. A few times, I half-fancied that I saw a few tall creatures running along with the truck a ways back in the VA swamp. When I slowed and looked harder, I saw nothing. Right before we left the GD swamp though, I stopped and looked back. I swear that I saw a dogman in the swamp, watching me leave the area. Ugh.

This experience has made me much more aware of my surroundings and I now look over my shoulder a lot and I examine the woods thoroughly at all times. I guess that's good and I thought that I was doing that, but I guess that I wasn't. I researched this phenomenon for a long time. It would seem that I may have opened myself up to the other side. I think that I might have brushed up against him again in the Blue Ridge Mountains at The Meadows of Dan and again in the Great Smokies.

After a while, i got brave enough to tell my story to my two best friends. They teased me a little bit, but knew me well enough to know that I was serious and then we all talked about it quite a bit. One of them spends a lot of time in the outdoors and his property backs up to a large wooded area on the outskirts of Lake Nockamixon, somewhat near Philadelphia. He hesitatingly started to tell me about some strange occurrences on his property. He said that he wanted to tell me about it now, due to the story that I just told him.

He thought that people would think that he was crazy, but now he felt comfortable enough to say something, since he thought that I might understand. His property is irregularly shaped and is about 20 acres. As I said, it is attached to a large park surrounding a reservoir. The area has a very long history and it was once sacred land to the Lenni-Lanape Tribe of Native Americans that lived here. In fact, there are stone structures nearby attributed to them and a nearby road is called Top rock Trail, though he would like to keep his actual address private.

So, he began to tell me that over the last month or so he started hearing strange noises at the back of his property. He did not have a large fence. It was just enough to mark the property really and to warn hikers of his dogs. His dogs are actually pretty friendly, unless you have bad intent.

So, these noises were snarling and kind of a baying sound... like wild dogs and the shed which he uses to process deer was damaged .twice It wasn't damage that you would expect from humans.

It was being clawed, stretched, and torn into. He repaired it with strong materials, but whatever was breaking through, got in again anyway. So he decided to watch from the house with binoculars and see what was disturbing the shed. He figured it was a bear. Bear were extremely rare in the county, but not completely unheard of. It could be wild dogs too, but he didn't think that they were strong enough to break his repair.

When he first started watching, nothing happened. One night, the surrounding forest noises just stopped and his dogs started acting strangely. They were usually alert and ready to go out. They had their heads down and tails between there legs and went down into the basement and refused to go out or come upstairs. He was watching the shed with the binoculars and he thought he saw a shadow move in the back. As he watched, (and felt very uneasy himself), he swore that he saw a wolf-like creature walking on 2 legs near the shed.

He shook his head to clear it and looked again. He said that the sight of this thing scared him more than anything he could remember and that he thought that maybe there was more than one.

He considered going out with a shotgun, but thought better of it. The next day, he went out to the shed and the dogs seemed normal. He saw no damage to the shed, but he found very large dog/wolf tracks and his dogs wouldn't go near them.

After he told me this story, I went out to his house and during the day, we looked at the shed. There were people in the woods walking around and I had no strange feelings. Everything seemed normal. I looked at the tracks, the shed, and saw his apprehension. That night, we built a fire and sat out in his back yard and watched the place. Though we saw nothing, there were a couple times during the night that we felt weird, like something was watching.

I told him of an idea that I had. I told him to start processing the deer off of the property. I knew it was a pain, but if he was going to continue living at that great house, in harmony with the surroundings, it might be better to avoid attracting attention, especially if a new animal had moved in. I told him to consider what he would do if a bear family was living in the woods. He would take precautions then, right? He said that he agreed and we washed the shed completely with a bleach solution and gave it a new non-food purpose.

After that, the visits and the activity stopped. I would not say that they completely disappeared, but they may have moved on. We never felt the presence again…anywhere in the park. I consider him extremely fortunate, because I felt in my heart, when I had my encounter, that the dogmen I saw, were part of, (attached), to the area…that they were going nowhere. Maybe it had to do with food and activity. I don't know what I would do if they followed me home!

It is strange how one can tell the difference between a BF, a predator, a human, and a dogman. Each are a completely different feeling, but the dogman holds a kind of "terror feeling," like the alligator, but much worse. Its almost like an instinctual fear. But…I never really saw it on those 2 trips of mine and I didn't see it on my friend's property, but my senses told me that it was the same feeling as in the Alligator National Park.

I have to say that I had that feeling a couple of times in rather small patches of woods and surprisingly, once in an urban woodland near my house in Philly. I know that sounds really creepy, but you know when you are being watched and I now know the feeling of the dogman watching now. I had a horrible feeling that maybe these things follow you over long distances, but I decided to dismiss that thought, for my own sanity.

We still love to go camping and be out in the woods, but as I said, I am always wary even if I don't tell the wife…as I want her to be relaxed and have a nice time. I am always armed with either .357 or a .45 and she is used to that. We tend to stay above the Shenandoah and we spend a lot of time in Vermont. We actually bought a house there and I would say that I have not seen or felt anything that far north. Past the boreal forest, there is bear and moose, but nothing else to be seen. Maybe it's the cold whether.

Chapter Five

Metal-Mouthed Dogman of Clare
~ Linda Godfrey~

Clare, Michigan, is known for several odd phenomena, and thanks to a recent eyewitness report, I'm about to add a third.

Set very near the center of Michigan's lower peninsula, Clare became the site of a string of UFO sightings in the late 60s and early 70s when strange lights were seen along Herrick Road. There have been UFO sightings about 14 miles south in Mount Pleasant, as well.

It was in those same decades when area was also known as home of the Clare Deer Man. Most witnesses described that creature as fairly human from the torso up, including a human-like head, with eyes that were between animal and human. Its lower body, however, featured the legs and hooves of a deer or goat. It sounded like a satyr; something torn from the ancient legends of Pan.

In 2005, however, a young woman driving home at night from her job at a nearby golf course at about 11:45 pm on Sunday, August 28, saw something entirely different.

The weather was clear and warm, although Tropical Storm Katrina was at that time pounding the nation's southernmost states. As she tooled along in her 1996 Dodge Intrepid, she was sober and alert to her surroundings. She wrote:

I had just turned onto Eberhart Rd. which is a dirt road from Beaverton Rd. I was traveling south on Eberhart, and going slow as I had just turned the corner, when I spotted the creature on the right hand side of the road, under a yard light that someone had placed at the end of their driveway.

At first I was unable to tell what it was, only that it was large and seemed out of place. In that area of the world we have many deer, and raccoons, rabbits, and the occasional coyote, wolf, mountain lion or bear. But even from a distance and with it hunched over, I could tell this was out of the ordinary.

I slowed my car to a stop as I approached the creature, at this point I still had not seen its face as it was hunched over eating the garbage that had been placed by the side of the road for pick up the following morning. The visibility was good, as the home had installed a yard light, at the junction of the road and their driveway, and the garbage and the creature were directly under it.

The first thing I noticed about the creature was that the fur was very short, like a horse's or a greyhound's, and shiny. The second thing I noticed was that it was in an odd position as it appeared to be hunched down with its arms and face low to the ground, as it was eating garbage, and it had shoulders like a man. It was jet black, and very and I cannot stress this enough, very heavily muscled, as it ate I could see the muscles in its shoulders and arms moving under the skin. It was larger than a man, perhaps the size of a bear when hunched over. If I had to guess, when fully standing, I would assume it to be over seven feet tall.

At this point my mind was going a million miles an hour trying to identify this creature, the fur and shape was like nothing I had ever seen. Then it looked up, and I still get goose bumps to this day when I remember that face. Its face was horrible, I mean really horrible. The eyes were red, like a sort of red that appeared to give off its own luminescence, this could be incorrect and it could have been that they were just incredibly reflective from the yard light, but he was looking at me, not up, so it was very odd to me that they should glow like that.

The teeth were the most predominate feature, they looked so unreal that I stared until it moved trying to convince myself that I was seeing it wrong and there was a logical explanation.

The face was shorter than most werewolf or dog man images, and the teeth were so large that it seemed impossible that such a thing could exist. It was hideous, and so terrifying. It had ears like a Doberman but they curled in slightly, and it was sitting back on its heals, and it was built like a man from the chest up, shoulders and all. Then it made a lunge of sorts towards the car, the last thing I saw before flooring it out of there, was the face twisting into a snarl and that it was coming towards me.

I literally floored my car, and it fish tailed on the gravel and I nearly lost control. I was so frightened, all I could think of was to get out of there and get out of there fast. I was living a couple of country blocks down the road at that time, and was so scared that I pulled my car up to the door and ran inside and locked the doors. I haven't seen it since, and outside of seeing a pair of ghostly legs crawl behind a chair while living in a haunted house, several years after seeing the dogman, I have never seen anything else that couldn't be logically explained.

When I told her about the UFO and deer man reports, she said she had never heard about them although she grew up only a block from Herrick Road. But they made a sort of strange sense to her. She replied,

I grew up with an intense fear of UFOs so intense that even as an adult I never felt safe in the area, yet oddly enough as soon as I moved, it was simply gone. Even as an adult, I could not sleep in the dark, and I would have these awful nightmares about getting sucked out my window, or being lined up in the back yard at night with the neighbors, no one being able to move or speak. *shudder* I left and moved to central NY in 2011 and have had no fear since.

Clare is also located about 20 miles east of Evart, known for glowing cemetery lights, and about 32 miles from Reed City, where several dogman sightings have taken place. The Manistee National Forest lies just westward of that town.

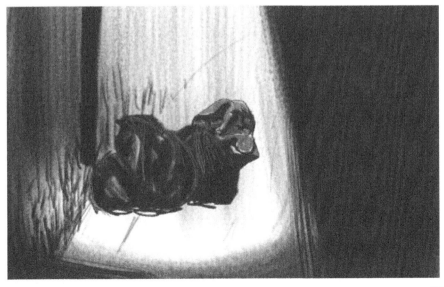

As I told the witness, what she saw didn't resemble the majority of dogman reports I receive in regard to its huge, muscular shoulders—canines don't have shoulders —its sleek, shiny black fur, and of course the horrific face and giant, metallic teeth. I believe dogmen with this description usually have some association with other unknown phenomena such as lights, UFOs and other strange creatures, and Clare certainly fills that bill.

These, black-furred, super-muscled monsters are not unique to Michigan but are reported from all over the world. In fact, I have another new report of a similar creature seen near Houston Texas. Watch for it here soon!

https://lindagodfrey.com

Chapter Six

THE SHENANGO VALLEY WEREWOLF
~Brian Seech~

The Shenango Valley, nestled in Mercer County, is a scenic wonder. The tranquil waters of the Shenango Lake belie the fact that a legendary creature may stalk the surrounding countryside. Strange sightings of an unknown animal began to occur around the mid 1950's and continue possibly through the present day. The local residents dubbed this creature the Shenango Valley Werewolf.

Margie was walking with her mother and several young friends, when suddenly from behind her she heard a howling noise. Quickly she turned around and stared in horror as the figure was running toward her on all four legs. When the creature reached her it stood up on its hind legs and grabbed her and started to pull her down the road. Her mother, who was standing several feet in front of her, grabbed Margie's hand and quickly yanked her away from the strange beast.

The creature quickly turned and ran back down the gravel road from where it came. The creature was again spotted several days later near an abandoned house at the end of the road where the initial sighting took place. This time the creature was seen to be wearing a blue shirt and pants. The creature, after being observed, quickly ran away into the woods.

The creature was dubbed "dog-boy" by the locals in the 1950's – 1970's. Sightings of the creature were rumored to have been seen throughout the area, sometimes it was been by the Shenango Lake, other times, loud howls were heard throughout the area.

Flash-forward to 2010. Our research group (CUE) Center for Unexplained Events had a vendor table at a local paranormal conference. The event dubbed MAPS CON (Mysteries and Paranormal Society) was nearing completion when Margie came over to our table and read the flyer "Have you seen this creature" on the flyer was a depiction of a canid looking werewolf creature. "I saw something strange like that," Margie stated. We went on to interview her and she explained her encounter. She described the creature as follows; the creature was dark brown, about four feet tall and hair covered over the entire body. One of its' arms appeared to be almost deformed, the hands seemed cloven. The face appeared flat. The one feature that stood out was the eyes, the eyes appeared to sag.

When I re-interviewed Margie in 2014, she sketched the creature (see drawing). During her sketching of the creature, she remembered the feelings of dread on seeing this creature. In June of 2014 we went to the location of the sighting in Jefferson Township, Mercer County, PA and photographed the sighting area. The questions remain what did she and all of the other eyewitnesses encounter, a deformed child with hypertrichosis? (abnormal amount of hair growth over the body – informally called "werewolf syndrome");

was it an unknown cryptid?; was it an interdimensional creature trapped in our realm and unable to travel back to its' own dimension? The phenomena then changed course in the 1990's with the next wave of sightings. The creature being observed was much bigger and its appearance was now more wolf-like. The Center for Unexplained Events interviewed an eyewitness (who we have now grown to know personally) about a series of encounters beginning in the spring of 1990.

Mark was walking with several friends near where he lived in Mercer County PA when they heard a noise at a nearby park. He shined his flashlight up towards the park when he was startled by what they observed. In the glow of the flashlight stood a huge, dog-like creature that was feeding on an animal, the light made it stand up on its hind legs and turned toward the boys. The creature looked like a man with a dog or more wolf-like face. It quickly ran into the woods, leaving a wet dog-like smell. The boys departed swiftly. The next day they found a deer that was eaten and broken branches about six feet off the ground.

The next encounter took place the following year in the fall of 1991. The same eyewitness came home from fishing and decided to look for his friends. He drove by an old water company in his neighborhood and found them. They appeared upset and they stated that they had just spotted the wolf-like creature again, it had just ran into the woods before Mark arrived at the water company. They drove by an old dirt road close to the water company. Mark shined the spotlight down the road and saw the creature standing by an old utility pole. The creature quickly departed after seeing the light shining in its face and fled back into the wood line. The next day the group went back to the spot looking for evidence. Although they didn't find any physical evidence, the group measured a spike in the utility pole where the creature was standing beside and it measured at 7 feet.

Eyewitness Drawing © C.U.E

The creature was described as wolf-like, covered in grayish-white hair and its eyes glowed red. The legend even dating back into the 1950's of the dog-boy seemed to center around the idea that a satanic cult or a coven of witches conjured up the creature to protect the land. Ironically, after an old red abandoned house (which was said to be where the creature was summoned from) was torn down and the sightings appeared to have stopped. What exactly was the Shenango Valley Werewolf? Did the sightings really stop or did people stop reporting them? A secondhand report was given to author Thomas White in the 2000's of a strange beast with dark fur allegedly encountered behind a home near Hermitage. The eyewitness still refuses to disclose any information about the sighting.

If you are in the Shenango Valley area at night and happen to hear a howl or something moving about in the forest, perhaps you too will encounter the Shenango Valley Werewolf.

C.U.E. & Center for Cryptozoological Studies

The Following are newspaper clipping of the investigation by the C.U.E.

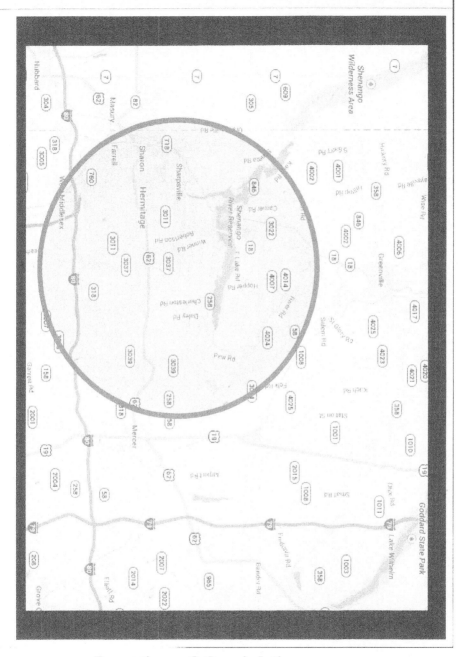

Location of the sighting area

Allied News

SERVING THE GROVE CITY, LAKEVIEW, MERCER, HARRISVILLE AND SLIPPERY ROCK AREAS

Since 1879

Destination America

A re-enactment of local Dog-Boy sightings, as well as first-person accounts near Mercer, can be seen at 9 p.m. Saturday on Destination America's "Monsters and Mysteries in America." Seech's next project centers on the Butler Gargoyle, a tall, winged-creature with a helmeted head that has been spotted at least 14 times in areas including Chicora. Info: www.destinationamerica.com

MERCER COUNTY

Hairy tale to air

Locals recount experiences with Dog-Boy

By Monica Pryts
Allied News Staff Writer

Two Mercer County residents were interviewed about their encounters with Dog-Boy for this Saturday's episode of "Monsters and Mysteries in America" on the Destination America channel.

Airing at 9 p.m., those tuning in will learn about local sightings of the werewolf-like creature, as told by Margie Lytle of Mercer.

Mark Headings of Sharon was also interviewed, but unfortunately his scenes had to be cut because of time and length restraints, according to a publicist for Discovery Communications early Tuesday afternoon.

Brian Seech, a former Grove City resident and co-founder of the Center for Cryptozoological Studies and the Center for Unexplained Events, will also be featured; he co-hosted "Strange Cryptids of the Ohio Valley" in September at the Community Library of the Shenango Valley, Sharon, and has worked closely with Lytle and Headings.

"This is gonna break it open," Seech said of his hopes for the show to shed more light on Dog-Boy, among other legends.

According to the show's website, "Monsters and Mysteries in America" goes
See **HAIRY**, *page A-2*

Allied News 2/4/2015 Grove City
By Monica Pryts

The Herald

LIFE: Tune in Saturday for tales of Dog-Boy encounters | A-9

FRIDAY February 6, 2015

Headings off the bed. He saw the creature hit the fence as it ran off; Headings was too scared to leave the camper and stayed there until morning.

He said he's heard stories over the years about a cult-like group at a nearby home that performs rituals, one of those acts possibly summoning Dog-Boy.

He shared his story with the TV crew and while they didn't have time to film the dirt road and wooded area, Headings, like Lytle, is relieved to be able to talk about what happened.

"I hope more people will open up about more things," he said.

Seech is hoping the same, and urges folks to report such findings to him. He got some new Bigfoot information from those who attended his library talk, which attracted about 100 people; he may hold another event this fall.

And for any skeptics, he suggests viewers keep an open mind, and Seech has become better at reading people who have a sighting to report.

"You can tell when people have had a genuine experience," he said.

He'd also like to see younger researchers pick up wherever he and his team leave off, whenever that may be.

Seech's next project centers on the Butler Gargoyle, a tall, winged-creature with a helmeted head that has been spotted at least 14 times in areas including Chicora.

"I'll always remain curious," he said.

Find Destination America on channel 103 on Time Warner Cable; channel 440 or HD 125 on Armstrong Cable; channel 286 on DirecTV; and channel 194 on DISH network. Info: www.destinationamerica.com

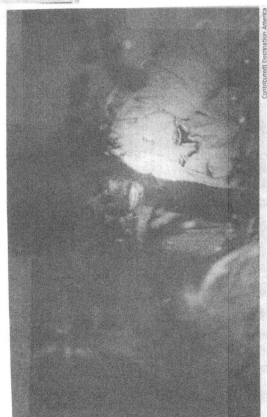

A re-enactment of local Dog-Boy sightings, as well as first-person accounts, can be seen at 9 p.m. Saturday on Destination America's "Monsters and Mysteries in America." Contributed/Destination America

The Herald 2/6/2015

2/6/15 THE SHARON HERALD

Tune in Saturday for stories of local 'Dog-Boy'

By MONICA PRYTS
Allied News Staff Writer

Two Mercer County residents were interviewed about their encounters with Dog-Boy for Saturday's episode of "Monsters and Mysteries in America" on the Destination America channel.

Airing at 9 p.m., those tuning in will learn about local sightings of the werewolf-like creature, as told by Margie Lytle of Mercer.

Mark Headings of Sharon was also interviewed, but his scenes had to be cut because of time and length restraints, according to a publicist for Discovery Communications early Tuesday afternoon.

Brian Seech, a former Grove City resident and co-founder of the Center for Cryptozoological Studies and the Center for Unexplained Events, will also be featured; he co-hosted "Strange Cryptids of the Ohio Valley" in September at the Community Library of the Shenango Valley, Sharon, and has worked closely with Lytle and Headings.

"This is gonna break it open," Seech said of his hopes for the show to shed more light on Dog-Boy, among other legends.

According to the show's website, "Monsters and Mysteries in America" goes straight to the source of such stories, talking to those who have experienced something that can't be easily explained, "local legends passed down generation to generation tell of monsters hunting nearby woodlands, ancient spirits walking amongst the living, and alien creatures paying earth a visit."

"Try to contain your screams," the website warns.

Saturday's episode focuses on the Florida Zombie, the Dybbuk Box, and "young girls are being hunted by the Shenango Dog-Boy."

"You can live in an area and not know what's going on. This is a good example of that," said Seech, who lives in Hopewell, Pa., with his wife Terrie, co-founder of CUE and CCS.

He's been on the show before, sharing his research about creatures like the Devil Monkey and the Michigan Dog-Man, but he's always excited to expose more people to stories he believes are becoming more commonplace because society has become more accepting.

Seech was filmed by the crew this past fall in Pittsburgh, providing notes, photos and other materials he collected during his meetings with Lytle and Headings. Everything went well, and he's excited to see it all come together, as are Lytle and Headings; they also attended the September library presentation.

Lytle was 8 or 9 when she had a run-in with Dog-Boy in 1955 or 1956 in Jefferson Township, off Hopper Road, not far from where she grew up on nearby Ballpark Road.

She's been telling the story for many years and first met Seech in 2009, but she's glad to have the chance to recount her experience for a large audience.

"It's just a relief that someone finally listened," she said of Seech.

She also met with the TV crew this fall. They interviewed her at home and Lytle took them to the abandoned farm where she saw Dog-Boy one night while out for a walk with her mother Mary and a group of neighborhood children.

"I don't have to ad lib any of that," Lytle said of what happened; she still has flashbacks.

As their group neared the property, a working farm at that time, Lytle's mother mentioned she heard Dog-Boy lived there, in the small milkhouse on the edge of the property. Lytle doesn't know who owned it then, or who owns the land now.

Several older children in the neighborhood had stories of unexplained, horrific howls; that night Lytle heard a howl and a screech, and her mother ordered everyone to run. The creature grabbed Lytle's left hand, her mother helping her break free, Lytle said.

"That was scary stuff," she said, recalling the large animal had a lot of dark hair; it returned to the milkhouse as she ran away.

None of the neighbor kids saw what Lytle believed to be Dog-Boy, but retelling the story has been therapeutic for her, and has helped her come to terms with what happened. She hopes others with similar stories realize they're not alone.

She plans to watch the show with a friend and is interested to see the re-enactment; she sent the network several childhood photos to help them with the casting.

"I don't know how I'm going to react to it," said Lytle, the youngest of six children.

encounters

She has another mystery on her mind after a recent visit to the abandoned farm; the barn and milkhouse no longer stand, apparently having been demolished.

"That was a landmark," she said, puzzling over what could have happened.

Lytle, Seech and Headings all expect a good response to the Dog-Boy tale; people will enjoy the local connections.

Headings spoke with Allied News Monday evening, recalling two times in the early 1980s when he crossed paths with a creature he's certain was Dog-Boy, and he wasn't the only one who saw something unusual.

One day he and his friends were walking down a dirt road off Charles Street in Hermitage, not far from Headings' Sharon home, when a few of their friends on bicycles came out of a wooded area, saying they had been chased by Dog-Boy.

About a week later, he was in that same area one night, armed with a flashlight, when something standing about 6 or 7 feet tall came down the dirt road.

"I saw the creature walk out," he said.

Another night, he slept outside his home in a pop-up camper, except he didn't get much sleep. He heard a noise that sounded like a dog breathing heavily outside. The animal then slammed into the side of the camper with enough force to knock

Continued 2/6/2015

Chapter Seven

Encounter with a Dogma
~Ronald L. Murphy Jr ~

This is a true story. It may seem like a work of fiction, strange and illogical in its structure and content, but I assure you these events actually happened. Like every good horror story, this one too has a monster. In this case it involves something known colloquially as the "Dogman." It is an ambiguous creature that is anthropomorphic in shape, but gives the initial impression of being somewhat canine. In short, the Dogman is an enigma. This is not a new cryptid in the paranormal universe—indeed, the notion of the Dogman goes back to at least the Pleistocene, if not further.

. It serves as the quintessential boogeyman, something immediately recognizable yet grotesquely brutish in shape and behavior. The Dogman is, in essence, the stuff of nightmares. The Dogman straddles two worlds, blurring the line between man and beast. Mankind has always been preoccupied with the notion of transformation, of moving from one state of being to another. It is part of human nature to marvel at the natural world and to fear it.

There is a healthy sense of reverence hardwired into humanity's psyche toward the forest and those things that go bump in the night. And that concept exists even today. This is why researchers, such as myself, set out into the woods and the remaining wild places and attempt to track down conundrums such as those known as the Dogman. We want to get at the truth that lies in the shadows. And I am sure that is why you are reading this book now.

I consider myself a cryptozoologist. I'm educated in various concentrations and experienced with the workings of the natural world and in the ordered chaos within the human imagination. Now, this doesn't mean that every time I venture out into the forest I find evidence of a monster. Indeed, I am blessed with such a healthy share of skepticism that I often go out not actually believing in the quarry I seek! However, I am also knowledgeable enough to know that something is going on out there and has been for a very long time. In the ancient tales told by indigenous tongues and in the modern eyewitness accounts, there are frighteningly similar reports of something like a Dogman.

. And this is why I still head out into the vast reaches of the pine forests and the swampy bottom lands. But because I am an academic researcher trained in the disciplines of anthropology and archaeology, most of my investigations are carried out in a library.

As I sifted through decades-old reports of hair covered, dog-like creatures prowling about on two legs, and as I reread eyewitness testimony of areas where such creatures have been seen, I pinpointed an area that seemed to be a prime location for an elusive creature such as a dogman to hunt without much human interference.

I ventured out around 9pm on a cold fall night. The night was hideously dark. My flashlight scanned the leaves that were orange and red, but still lingered on the trees. I was well off the beaten path, and any nuisance of light pollution was obscured by October's cloud cover. But this location was the rumored haunt of the Dogman, so I zippered my coat against the chill and set out down a path that twisted under a hill on which an abandoned cemetery glowered forlornly through the rustling tree limbs. Dogman is a cryptid often associated with the abode of the dead, with a lineage stretching back to Anubis and its ubiquitous presence in Egyptian funerary rights. However, this creature is also frequently seen even to this very day prowling around Indian burial mounds and on grave sites in Native American reservations. From the point of folklore, this area was the ideal site for this creature to be found.

But our dogman has always been a creature of the liminal regions, that ephemeral threshold between this world and the world of the "other."

." And out here, on this sullen trail that carved its way through the Autumn forest, I felt as if I were hopelessly adrift in this space between worlds.

As I swept the woods with my night vision camera, I was struck with the incredible feeling that I wasn't alone out here. I saw no animas through my eyepiece. No squirrel rustled through the leaves, no rabbit stirred the stillness. No. It was something else. Light anomalies seemed to follow me along the darkened path, nearly indiscernible in the Fall night. But I could plainly see the pinpoints of unmistakably living light flitter through my lens, little soft glows that darted and danced in the twisted undergrowth that lined the path, occasionally skipping out of the tangles and illuminating the trail. Even after my eyes adjusted to the night, I could see these illuminations without electronic aid. But I ventured onward, following the lights that moved forward, leading me further field.

Then I came to the long abandoned railroad bridge that spanned a river that flowed lugubriously like a black ribbon in the valley. The little lights that had accompanied me now left me as the woods halted at the bridge. In the center of the bridge, I felt utterly exposed. If I were being hunted, I was now vulnerable. There was nowhere to run and, in the stark nakedness of the bridge, there was certainly no place to hide.

The rather soothing electronic glow of the camcorder which I carried suddenly seemed to have its energy inexplicably drained, the battery light suddenly flicking on. Above me the sky that enclosed the world in its inky embrace seemed to spark with a static electricity, reminiscent of the electricity produced when you pull blankets from the bed. Something was happening to the world around me. Suddenly, a visceral urge shuddered my body. I knew instinctually that I had to get out of where I was. I must get back into those confining woods and make my way to my car, parked nearly a mile and a half away. At this point it was not a choice, it was an imperative drive to stay alive.

I turned. Behind me the trail disappeared as it was swallowed by the gaping void of the woods. But immediately as I turned my eyes widened in trepidation. There, where the trail plunged into the interminable abyss of the forest's gullet, a light of living intensity sparked to life, shimmering and sputtering like a contained Roman candle suspended in the air. I immediately called out, knowing rationally that there must be a person behind it. My mind worked with the information before me, struggling to process it intellectually.

It must be a person. What else could it be? I glanced into my night vision camera's eyepiece. There was nothing there. The light was gone as quickly as it came.

And there was no person on the trail. Unnervingly, my battery icon in my camera was flashing an urgent SOS, begging to be recharged. It was as if the light that manifested before me used nearly all the remaining battery in my camera to power itself to life. Empirically that made no logical sense. However, cold and shivering from uneasiness, nothing else made better sense. And that gnawing instinct that sent a shiver of compulsion through me to run for my life was much more forceful now. I knew I must get out of these woods.

I quickly walked the bridge down towards where the light had ignited. Soon I made it to the tree line, shrouded by the overhanging branches that closed in on either side. I was alone. Utterly. The night was empty. Then an electricity shivered my body that instilled fear into me. I felt as if I were being watched. I also had the startling realization that I was being hunted. My quick walk became a trot until I found myself fumbling into the darkness at such a pace that my heartbeat pounded in my ears. My face stung as branches slapped my cheeks, my feet tripping over roots and through brambles as I veered from the trail in the absolute darkness. Soon I realized I would inevitably stumble and fall. Every primal instinct in my civilized brain screamed, insuring me that is I fell I would be pounced upon. Something in the night was stalking me.

Then something inexplicable happened. Even telling you this now it seems rather difficult to believe, but I assure you of the veracity of the event. Off to my left, out in the woods and higher up, as if a person called out from the cemetery on the hill, I heard my name called. "Ron," is said. The voice was androgynous and undefinable. It was a quick gasp, yet stark in its immediacy. And it was an urgent call. Instantly, and hard to my left, something stirred in the twisted brush and in the littered leaves that covered the forest floor. On the other side of the trail, something was moving closer, stalking. I could hear it breath, the night air wet in its mouth. As the sticks snapped under its weight, I realized with unmitigated terror that I could "see" what was there. It was as if the synapses in my brain fired to produce an image of the creature that I knew was part of my collective unconsciousness.

It was primal and bestial. It was the same menace that had pursued the human species since we first walked upright out of the African savannah. It was a dogman. My brain constructed the creature, drawing it out of the memories locked within my DNA. It was huge, hulking. I could see each step it took was on paws that were more like hands. Its snout was lined with rows of needle-like teeth.

Its nose smelled the air, pulling my scent out of the cacophony of odors that filled the woods. And those eyes. They were the eyes of a nightmare because they were the cause of nightmares since our genesis. The red eyes, narrowed with acuity. Seeking me out.

But I knew it was there. And it knew it as well. The little voice that called me name alerted me not only to the direction of the path that led to my car, it also shook me to attention. The Dogman may be a part of our evolutionary journey, a clever, self-aware creature that still stalks the civilized world, preying upon us, but we have an intelligence that lifted us out of the jungle. I back-peddled, looking through the viewfinder of my night vision camera toward where the creature stalked in the darkness.

The battery, which had been dangerously low, now was reading that it still had over half its power. I quickly spun the camera around, finding the trail in the blackness of the night. Then I slowly followed the trail, keeping the night vision trained on where the Dogman lurked. I could hear it snarling, its breath wet and anxious. But it never let its hiding place. It remained concealed in the thicket. Even when I had finally made it to my car and locked the doors before turning the key, I knew the creature was still crouched over in the forest, its eyes red, scanning the path, waiting for prey.

In reviewing the video evidence, I could see the little points of light were captured. So too was the light that ignited so brilliantly when I was out on the bridge.

I realized that it was that light that drew me back into the woods. Why? Did the light have an intelligence? Did it know that there was a creature hunting in the woods? Other reports associated with the Dogman have occasionally reported strange lights. Is the Dogman interdimensional, and are the lights the visual evidence of portals opening? I'm not sure. But the lights did coincide with the appearance of this creature. And what about that voice that called my name? It did not appear on the film. That was my own little experience that cannot be shared. It was personal and poignant. The sounds of the creature prowling in the brush can be heard, but without that voice I am convinced I would have been pounced on from behind, blindsided by this predator. I feel that I would have been another missing person's case without this voice. But this time I made it out of the woods. I had encountered the dogman. And I survived.

Chapter Eight

Wolf man
~Danielle Steadman ~

My name is Danielle Steadman and I'm from North Carolina near Charlotte. I'm 42 yrs old & a proud single mother of 4. I grew up on a 200 + acre farm as a child where we grew our own vegetables & fruits and raised & killed our own pigs, chickens & beef cattle. I also come from a large family of hunters/fishermen and was taught to shoot, hunt & fish at a very early age. So I know my way about the woods & around weapons fairly well... Ok understatement probably. I have always been able to out shoot most people & my hunting skills are way above average. I can break my guns down & clean them and skin out & butcher a deer, hog, cow or anything else. I know all the types of animals we have around these parts (& around the world for that matter)... I want you to know this so you realize that what happened to me was not a case of mistaken identity.

I'm also one half Native American (Cherokee) and was raised hearing stories from elders about other things most people don't believe are real in this world. Yes, some of them being Sasquatch & Dogmen.

It was Sept or Oct of 1987 & I was 14yrs old. I was always mature for my age (I was an only child with a very sick Mother at the time who died unfortunately a yr after this incident) but I also looked mature (older) and had a boyfriend who was 16 & able to drive although I wasn't allowed to (as my Momma called it) car date with him. So on this particular night I snuck out of the house to meet him & 2 other friends (my friend Jill & her boyfriend) around midnight. I've changed everyone's real names to tell my story as I'm not in contact w these people anymore.

I remember that the moon was pretty full this night because of how it illuminated Jill's boyfriends dirt road when my boyfriend parked his 69 Camaro on it. My boyfriend's car was really loud so he had to park it at the beginning of that dirt road so we wouldn't wake Jill's boyfriends parents then walk about a mile down the dirt road to reach Jill's boyfriends house to let him know we were there. We had already picked up Jill so he left Jill & I in the car while he walked the mile to the house.

My boyfriend wasn't gone long maybe ten minutes and Jill was relaying to me all the school gossip as we listened to the radio, when I started to smell this Gawd awful stench... I'll NEVER forget that horrendous smell. It smelled of rotten meat once maggots have set it, a musky pungent scent, wet dog & blood. Yes you could literally smell the metallic iron like smell of blood... I knew it well from slaughtering our own animals & hunting.

Jill was still jabbering away (she was quite a talker) when I interrupted her asking if she smelled that? At first she said smell what? And I was like you don't smell that awful smell? She took a good whiff & said eww yes what is that? I said I don't know but I don't remember smelling it when we first pulled up here. It was so bad we had to pull the fronts of our shirts up to cover our nose & mouth. Jill was like I guess the scent just didn't reach us right off the bat. I didn't know how but figured she must be right.

Jill went about talking & not even five minutes later there was a suddle nudge of the car. Like someone gently rocked the rear end. Jill shut up immediately when this happened, her eyes grew wide as she asked what was that. My heart had started beating fast and I said I don't know.

Then I smiled & whispered I bet it's the guys trying to scare us. It had now been long enough that if the guys had hurried they could've been back. And they were always pulling pranks like that.

We looked out the cars windows but couldn't see much because the windows had fogged up from us sitting in there talking for so long & I wasn't about to wipe them w my bare hand cause my boyfriend would have a fit over the finger prints (this car was his baby). Jill said ignore it, so we did & she went back to talking. And again there was a slight rock of the car. We both smiled at one another knowingly & Jill kept on talking...

But then the car was shook really forcefully. This time we laughed out loud & I yelled out to the guys that it wasn't gonna work. Another forceful shake of the car and again I yelled out that they might as well stop it wasn't working... Jill chimed in with a Nope not goanna work too. Still no answer from the guys. I yelled their names... No reply.

Jill and I were still smiling when the car was shook again but this time from the front and this time it was down right violent... We strained to see out the windshield but couldn't see anything... I yelled for them to cut it out and come on but still no answer... Now my smile was disappearing along w Jill's & I was starting to get alittle scared. Again a violent shake and I screamed stop it your scaring us... Nothing. I cut the radio all the way down, almost whispering "Guys?" Then I heard it... A low, guttural, deep, sinister growl...

The type of growl that resonates in your chest & sends cold chills up & down your spine!

The type of growl that resonates in your chest & sends cold chills up & down your spine! The hair on the nap of my neck stood on end! And I don't know how to explain it but it was no ordinary growl. It sounded canine like but there was something troubling about it. It was so ominous... so disturbing! To this day I can't quite put my finger on it.

What was that? I looked back at Jill and it looked like all the blood had drained from her face. Then the eerie screeching of what sounded like finger nails running from the front left quarter panel on the drivers side all the way down to the door handle. My eyes followed the slow, terrible sound in utter dismay ... And I knew at this point that this was definitely not my boyfriend because he'd never do that or allow that to be done to his car. Then it stopped. Jill & I were frozen, I don't think I had even taken a breath as all of this unfolded before my eyes. But then I watched in horror as I heard someone or something trying to lift the door handle. At that point I scrambled to jump the short distance across the driver's seat to lock the door!

This put my head & eyes maybe one & a half inches from the glass. Thank God steamy fog clouded the window but I just knew someone or something was on the other side watching me. I could feel the eyes boring into me causing the hair on the back of my neck to continue to stand. It was absolutely unnerving! I felt a type of fear that I had never felt before & pray I never feel again... It was a primal fear. A fear that tells you that your life is truly in danger. A fear where you feel like your heart is literally goanna stop beating because it's beating so fast that it's hurting. You feel sick, dizzy & a clammy cold sweat envelopes you leaving you with freezing goose flesh all over your body.

I was literally frozen with fear as I looked out that foggy window. I was so close that my breath was creating more steam on it and although I was terrified apart of me just had to see out... I think because the logical part of my brain was telling me that I was over reacting and that it was nothing more than a terrible prank the guys were playing or at worst a large angry dog. I even wondered if a wild hog had somehow gotten down outta the mountains and into our area as they can be quite large & make some awful sounds. Maybe I had even imagined the door handle lifting...

Slowly I reached up with a shaking hand & wiped a little streak of fog away from the glass.... To my ultimate surprise I saw what looked like a large dark canine type muzzle & wet nose pressed against the glass that snorted out another low, horrible growl... The breath from the creature created its own steam on the outside of the window. And then it looked like it rose up! Yes, the freaking thing stood!!! Stood up, as in on 2 feet - bipedal!!! Reeling from terror I slammed myself back into my own seat as Jill screamed what? What is it? My mind was racing...

Did I just see a dog stand!?!? And that looked like the muzzle of a German Shepherd or wolf??!! But it was HUGE and longer?!?! More narrow at the end?!!? Idk something wasn't right!!??!! No, no, no... surely not!!! Jill repeated herself snapping me a little back into reality but before I could answer her (which I didn't know how I would've answered her anyways) the entire car began shaking like this thing was standing there shaking it by gripping the top of the car near the upper part of the window & roof... Jill began to scream & cry. Which seemed to make this abomination madder and it shook the car harder, growling ferociously!

I was telling Jill to hush and be quiet because she was making it worse but she wasn't listening at this point.

I slid down into the passenger side floor board and was whispering to Jill to get down in the same position I was but she had crumbled up into a fetal position in the back seat where she was sobbing. I truly thought dear God we're gonna die & I was just waiting for this thing to come crashing thru the window at any second. I was hushing Jill this whole time & she did seem to attempt to calm down some. Suddenly the car quit shaking and the growling stopped. We knew the creature was just standing there... I knew it was listening for us & I very quietly urged Jill to calm down some more. Lying to her by saying that it was goanna be alright & to see the shaking had stopped.

Somehow Jill got her breathing under better control and quit sobbing. It seemed like the silence was deafening & much worse than the shaking & growling had been. After what seemed like forever we heard the creature move away taking a couple steps back from the car. You could hear it sniffing around the car window still tho & I'm sure into the night air. We couldn't really see it just an outline of it here & there as it moved on TWO legs around to the back of the car, sniffing away the entire time. It was incredibly tall and wide by its silhouette! I'm guessing over 7 - 8ft in height!!

The gravel was crackling & popping from the weight of this monster as it straightened and moved along the car!!! It was down right weird!!!

Then you heard it collapse down on all fours with a heavy thud... (now I know the popping & crackling was probably from it standing upright but at the time I thought it was the sound of the gravel beneath it's feet). It was obvious that the thing was very heavy and I'd guess 600 to 700 lbs maybe even more (considering muscle weighs more than fat... and I believe people under estimate their weight... anyways....). Thankfully Jill was still being fairly quiet. Why I had remained so calm was beyond me... Maybe because I felt like I had no other choice because Jill had been falling apart... Maybe because my Daddy had raised me up to be so rough & tough... I was an only child and although I was pretty spoiled by my folks I was also a hardcore tomboy from being raised on a farm and seeing animals die on a regular bases. I don't know how I had kept my wits about me during all this but somehow I had so far.

I heard the creature quit sniffing and it sounded like it was messing with the trunk lock. I was thinking oh my God its trying to get in here. And at that moment I remembered my door was still unlocked so with an unsteady hand I reached up, took a deep breath and slowly pushed the lock down hoping the thing wouldn't hear me do it.

But as soon as the lock engaged the locking mechanism it made what seemed like the loudest clicking noise ever and suddenly the dang creature quit fiddling with the trunk, paused just long enough for me to whisper oh shit then it ran, yes ran, around to my side of the car and was now right outside of my door again.... And I could hear it sniffing at the handle. I truly thought my heart was goanna beat out of my chest and I thought I was going to get physically sick!!

I sunk lower into the passenger side floor board and tears filled my eyes. I was absolutely terrified!!! So terrified, that to this day, its beyond my ability to describe just how extremely terrified I was at that time. All I could do was pray that the freaking thing wouldn't bust thru the window. Yet that irrational voice inside my head was going you know its goanna bust in here at any moment and get you!

Suddenly the sniffing quit and again dead silence. Oh hell what's it doing now I thought. And like to answer me I felt & heard the door handle lift. I was already in a very awkward position but I turned my upper body around so I could grab the inside handle with both hands just in case it somehow did get it open.

I knew that by the size of the thing I wouldn't have a chance against it in a tug of war over that door but I sure wasn't going to just let it in without some kind of fight!!!

Jill was starting to cry again and this time I ordered her to shut up. I'm pretty sure the creature heard me and it seemed to let go of the door handle. A few seconds passed and just as Jill said Danielle? the freaking thing started growling that sinister growl again & then it was back to trying to lift the door handle again but this time it was jerking at it hard enough to shake the entire car again. I was gripping the handle so tight my knuckles were white and my hands were really hurting but I didn't care. I wasn't letting go of that door!!!

Jill began to panic again... She cried out go away!!! And I lost it at this point and started screaming stop it, just stop it & begging it to go away too. I just kept screaming this over & over!!! By now I was crying hysterically too. And unbelievably it suddenly just stopped.

No shaking, no lifting, no growling just the sound of Jill & I heavy breathing between wet sobs. A minute passed and Jill asked me if it was gone? I didn't hear it leave but we had been crying & yelling and although I definitely did not want to I knew I was goanna have to look out the window.

Keeping one hand on the door handle, I scooted up just far enough to reach the glass but still kept the majority of myself in the floor board, hidden. And using my other hand, which had practically fallen asleep from the numbing pain of gripping that handle so tight, I managed to swipe a little space of fog from the glass. Slowly with eyes wide from fear I peeked my head up just enough to look out of the little space that I had made...

And to my horror, as my heart sank and my blood ran cold, I now found myself looking into 2 of the most deadly eyes I have ever seen. Its eyes were a freakish ember yellow and it just glared at me as it began that ominous deep growl again, obviously angry that it couldn't get to me! And this freaking thing was ALOT larger than I had estimated!!! It was enormous!!! Its head was the size of a grizzly's! In fact I'd bet a little larger!!!

I wanted to look away but couldn't... My eyes were looking at it but my brain just could not fathom the message my eyes were sending it!!! This was a monster!!! A real & true creature of the night!!! Something I'd over heard about in hushed stories the elders would tell!!! Something horror stories and legend's were made of!!! But it was supposed to be exactly that!! A legend, a myth, something that wasn't really real!!! Something made up!!! Fictional!!!

But oh dear God in heaven above it really did exist!!! And in that one moment everything I thought about this world was irrevocably and forever changed!!! Everything I thought I knew - changed!!! Everything I believed & didn't believe - changed!!!

Along with that monstrous head were tall pointy ears with long tuffs of hair coming from the tips & inside the ears. Its coat looked longer than a German Shepherds but just as coarse. It had tiny frays of white around its eyes and snout, like it was just beginning to age maybe. And it had a few scars... One above it right eye and one across its snout... From fighting perhaps... Or from killing! Then it just had to go and sneer at me, unfortunately, and I caught a glimpse of its huge ivory teeth! (My stupid mind couldn't help but interject "all the better to eat you with!" I guess my mind's way of trying to cope with this situation.) It was like looking at a very, very over sized wolf but there was also something very non wolf like about it...

Something unnatural... Something dangerous & very predatory. I knew I was looking at a true predator. A top of the food chain alpha predator!!! A real natural born killer!!! And my mind reeled as it screamed its a f-ing werewolf you idiot!!! Your looking at a werewolf!!!

Suddenly it jerked its head down the road looking in the direction that our boyfriend's would be coming from at any minute. I suddenly realized our boyfriend's would be back and that they were goanna be outside with this freaking thing. I began to feel even sicker & started to cry again because I had no idea how to warn them. And I just knew this thing was going to kill them & we were goanna have to witness the savage slaying just to be next!!!

Still looking down the road the creature lifted its snout into the air & began sniffing at the night air, its ears twitching in the same direction it was sniffing. And then it quickly snapped its head back in my direction, gave me one final terrorizing glare and charged off on all fours in the opposite direction of where our boyfriend's would be coming from. And boy could it move. It was lightening fast.

Moments later we saw two flashlight beams and heard the guys laughing and talking. Still paralyzed with fear neither Jill nor I moved. Tears just ran down my cheeks. The guys had to tap on the glass of the windows and say open up before I budged. Once in I quickly locked my door then leapt over my boyfriend and locked his door.

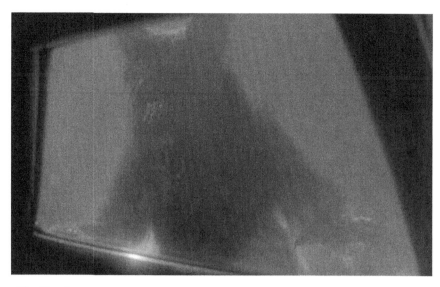
The Howling move

He was like what the hell is wrong with you and I could hear Jill's boyfriend asking her why she was crying... Then my boyfriend looked at me and said are you crying too? What happened?

Before we could answer them we all heard this terrifying eerie howl that cut thru the night like a razor blade. It was long... Almost sad but I knew that the creature making that howl wasn't at all sad... Perhaps mad that it didn't get its midnight snack.

Suddenly Jill & I screamed at my boyfriend "Lets go!!!" He looked at me and dared to say not before you tell me what happened! Jill & I were exasperated and we both began to cry harder as we nervously looked around

Jill whined to her boyfriend that she wanted to leave but I reached over grabbing the keys in the ignition to crank the stupid car myself and screamed at the top of my lungs "We have to go NOW!!!"

Shocked but with little reluctance my boyfriend cranked the car up and we sped away from the area where the monster had just been.

The guys were really mad that we wanted to go home immediately and I don't think they really understood what either of us was trying to tell them until later on. Of course we were both talking frantically, telling what had to seem like a very irrational story and I'm sure we weren't making much sense.

But the next day when my boyfriend came by there in the dust, that covered his car from the dirt roads were very large hand prints on the roof, windows, & wheel-well of his car... And all down the driver's side from the front quarter panel to the door... deep claw marks!!

My friends and I never spoke of the incident again and over time & years slowly lost touch. But as I stated earlier that incident changed me forever! And to this day I have trouble sleeping & will turn every light on around my house if I hear something I can't identify or if I get an overwhelming feeling of fear

Sometimes I have to take sleeping pills to sleep and anxiety medication for the panic attacks.

I watch my surroundings, especially at night. And I don't hunt or camp alone. I don't camp often at all. I'm very cautious and I've always kept big dogs... Rottweiler's. In fact this is the first time I haven't had a dog in many years but I plan on changing that soon.

I always carry a gun or have one close by. Although I doubt a gun could stop what I saw that night if it truly wanted you. But I'd rather have it than not. My ex-husband, who was one of the very few people I trusted my story with, even had me a silver bullet made. No I don't believe what I encountered was the mythical werewolf of Hollywood so no I don't think silver would kill it but it did make me feel better and I know my ex did that so I'd know he really did believe me.

I've always been accused of being paranoid and afraid of the dark. And I guess I am... I tell people I'm not afraid of the dark but afraid of WHAT'S IN the dark. And I have good reason to be. I don't isolate myself by any means but I don't do things I consider stupid either... Like walk down a desolate road at night alone or hear a strange noise and grab a gun to go investigate alone.

And for years I tried to suppress it and pretend it didn't really happen but I think it made things worse. When I finally got old enough to finally say ok this really did happen and its OK to admit it and believe it I started feeling a little better. Although I still feel very alone in the matter. But before finding your show I was finding ways to cope with it better... Like I started writing this fictional story about a werewolf.

Unfortunately this isn't something you can share with other people, even people very close to you usually. Most people are going to laugh at you and ridicule you. Saying you imagined it or it was probably some type of animal and you were so young and scared that you misidentified it. Trust me when I tell you that there is NO WAY to misidentify this creature. So for me finding your podcast has been a God sent!

Hearing other's say its true and they've seen it too helps tremendously!!! I hope to find some friends thru your show so I have people to talk to if I start having a hard time with it and sometimes I do. The nightmares can be overwhelming. And I suffer with alot of anxiety over my incident. I also have so many unanswered questions... Like what are they really? How many exist? Are they in packs or loner's? And why me?

What I don't understand the most though is why that creature didn't break the glass that night? It was intelligent enough to lift a car door handle & fiddle w the trunk but it didn't think to break the glass??? I've asked myself that question hundreds of times.

And why did it just run away like that? If could've had us all for a midnight snack???

I hope thru your show, talking to others, researching and writing I can learn more about these creatures. If anything I hope it'll help me to cope better with my situation.

Well you too have now heard my story and whether you believe it or not is up to you. Unfortunately for me I have no choice but to believe it and deal with it in some way, shape or form everyday for the rest of my life. Dogmen are a part of my reality whether I like it or not.

Chapter Nine

Something Unexplainable
~Donna Fink~

Here's the story of my experiences with something find unexplainable.

I heard of Bigfoot being seen twice in my area since 2006 in my local newspaper. I began contacting people in the area and found that a homeowner had witnessed something that sounds like a Bigfoot standing in her yard about 10:30 pm in 2011. This is what started my investigation in this area.

While looking for Bigfoot in the woods near the eyewitnesses' home, we would always search the area outside of the woods for footprints because it was easier to find prints in the dirt surrounding the woods rather than in the woods where there was so much leaf litter and brush. We found many prints that we believe to be from a Bigfoot and cast quite a few.

One day, as we entered these same woods, there was an 8-10 ft. diameter of wetter ground in a small hollow. We checked that area thoroughly before going into the woods and there were no prints of any kind in the hollow except for some tire tracks from the neighbor. Keep in mind these woods are only as big as 1 square block. I was standing in the back middle of the woods taking pictures of the large (20 ft.) Bigfoot shelter that we had found in those woods. My friend (I'll call her Melissa) was at the front left of the woods. Suddenly, I heard a branch crack and then a loud rhythmic electronic transformer type buzzing sound started in the back right of the woods. I thought to myself that someone must be using some type of equipment in the fields surrounding the woods or something. These woods are surrounded by cornfields on 3 sides and a house on the 4th side. The noise stopped but started up again in the front middle of the woods. I called out to Melissa to come back over by me and told her about the buzzing sounds I was hearing as she said she hadn't heard anything unusual. We decided to move forward to the front of the woods & look around.

I was still thinking someone must be working in the fields. Suddenly, we came across a 10 ft. clearing in the woods which was encircled by trees that formed a 50 ft. tall teepee. Melissa was standing in the middle of the clearing and I was outside of the circle. We started hearing the same electronic type buzzing sound coming from Melissa's backpack but the sound was much, much quieter. Melissa opened her backpack to see what was making the noise and couldn't see anything. Suddenly the noise came out of her backpack, circled around and around her until it was over her head and then 15 feet up and then gone. We couldn't see anything making the noise & became very frightened. Our first thoughts were that it must be aliens or something scanning Melissa and we expected to see a UFO coming any second. Melissa started to run past me out of the woods into the cornfield.

I grabbed her & told her that if she left me there and the aliens picked me up that I would make them land & pick her up, too. I said, "Besides, if you go out into the cornfield, you're only going to make it easier for them to get you." Melissa stopped then & we decided it would be best to leave the woods at that point. We went out the way we came in only to find 2 large (5 1/2 inches in diameter) wolf footprints in the muddy hollow that weren't there when we entered the woods an hour before.

It appears to be a front and rear print but how could a wolf leave only 2 prints in an 8-10 ft. diameter muddy hollow? Where were the other prints? We took pictures and measured the prints. We typically don't have wolves in our area as we are very far south in Wisconsin, only 1/2 hour from the Illinois border. Of course, there's always the possibility of a lone wolf making its way this far south, I suppose.

That night at approximately 2 am, we got a report from a young lady who called and said that she saw an "oversized" wolf with glowing blue eyes sitting along the roadside about 2 blocks from where we found the prints.

The next weekend, it was a beautiful sunny afternoon and Melissa and I were on the county highway heading for our Bigfoot area to do further research. We were traveling 55 mph going east. Along the opposite shoul of the road & walking west was what at first appeared to be a younger man. What we noticed first was the horrid condition of his hair. His hair was long and matted all over with the back of his hair matted into a ball at his shoulders. His hair made him look as though he had been living under a bridge under a pile of leaves for at least a year! The weird thing was, his clothes were perfectly clean. He had on a dark hoodie, jeans, blue canvas Kids type slip on shoes and a red plaid shirt. It was when he looked up at us as we were passing him that we both screamed at the same time.

His face was shocking. He had a bit of a heavier brow, dark circles around his eyes, a normal human nose, but a protruding wide mouth showing all of his perfectly shaped, very large, flat, straight teeth, both the uppers and lowers. Melissa said that she saw fangs. I didn't see fangs so I can't say for sure and that's because I was driving and trying to watch the road.

Even more shocking was that his teeth were yellow and streaked with green and brown, something we could both see in spite of traveling at 55 mph! And, the space between his nose and lips was much larger than a human's. His protruding mouth & teeth reminded me of a chimpanzee's. When he looked directly at us, I got the feeling he was telling me with a smirk that, "I know you and you know me." Well, we decided that we had to get a video and pictures of this guy so I made a U-turn as quickly as possible, came back around & passed him about a block ahead and pulled in by the cemetery, turned around, aimed the truck towards the road, got our cameras ready and waited from him to come walking past.

I didn't want to pull up next to him on the highway because I was almost afraid that he would shapeshift, turn into a wolf and jump in our window or the back of the truck. I know that sounds silly but his appearance put him into the realm of the unknown for us.

All of a sudden we see an unmarked, brand new white van with a raised roof heading east, the opposite direction of where what we were now calling the "Were boy" was headed. Melissa said, "They are going to pick him up!" I said, "They aren't going to pick him up! Who would pick him up? No one will!"

But the van made a U-turn & pulled up next to Were boy. The driver got out, pulled open the cargo doors and either threw Were boy in or he got in on his own. We couldn't see that part because there is a large tree in the cemetery that was in the way of the doors. The van then made another U-turn and headed back in the direction that it was coming, which was the opposite direction that Were boy had been heading. We pulled out within 10 seconds of the van in an attempt to find out what was going on and we could never find the van again in spite of driving 70 mph, looking into every driveway and crossroad along the way until we arrived in the next town 10 miles away. We drove and searched all through that small town trying to find that van and never saw it again. How is that possible?

We found plenty of large wolf prints in our research area and it appeared that once it was on 2 feet and chasing a deer. We know these were not dog prints because 2 very large, heavy dogs lived in the area and we measured their footprints and took pictures of their feet. Their prints were only 2 1/2 inches in diameter at most and didn't look anything like the wolf prints we were finding. It looked like something was trying to get the deer but we found no blood or anything. One day we were in the woods and heard something take 2 to 3 steps, the brush rustled a bit and then there was quiet. We knew that something was in the woods with us but it was daylight and we were trying to find it. It got to be almost dusk and we exited the woods.

We had parked the truck about 2 blocks down in another cornfield and were headed out of the woods and into the cornfield when we suddenly heard a very loud wolf howl with sort of a hyena/human laugh at the end. Melissa and I broke into a run for the truck down the rows of corn that was over our heads. I was so scared, I yelled at Melissa to shine the flashlight behind me and she asked me why. I told her that I wanted to know if a Dogman was coming after us because if it was, I was going to run faster. Nothing was coming out of the woods after us, thankfully. I think the creature was messing with us and just trying to scare us.

The next odd thing that happened was my husband and I were driving to our home, which is about 9 miles from the area where we saw the Were boy. We were on a country road within 1/2 a mile of the city limits of our home and it appeared that a deer was coming up from the shoulder of the road & going to cross in front of us. My husband had to come to a complete stop because instead of a deer, it was a large wolf, and it crossed the road in front of us within 2 feet of our car.

The wolf just trotted across the road, glancing around, but never looking directly at me, to my relief. The closest I can come to what kind of wolf it was is that it appeared to be a brown timber wolf. It was twice as tall as a coyote and definitely not a coyote. Is it possible that Were boy was just letting me know that he knows where I live and trying to scare me? I don't know the answer to that but it crossed my mind.

We did investigate 2 other reported Dogman sightings, one was within 2 miles of my home and the other within 10 miles.

We continued to find and cast wolf prints in our Bigfoot area off and on for at least another year and then didn't find any more wolf prints until February of this year 2015. The tracks appeared to be a wolf walking on all fours from the cornfield directly into the Bigfoot Woods.

I haven't been afraid to continue researching my Bigfoot area by myself in spite of what could be a Dogman in the area. I think that if a Dogman wanted to harm me, he had plenty of opportunity to do so. I don't go into the woods at night alone though. I also continue to search the woods where Travis Lund and his friends had their Dogman encounter.

There was only 1 time that I got a feeling of fear and that I needed to get out of the area right away. I was in by the marshes where Travis had his sighting and where I have been many times before, but this time I felt that something was present in some brush and trees although I could see nothing, I immediately left the area and didn't go back for at least a month or more.

Chapter Ten

"What Am I Seeing"
~KBRO~

Location: New Salisbury, Indiana

Contrary to what some may believe, this report was NOT submitted because it's October. A few years ago a parent at the school in which I (Charlie) teach informed me about her co-worker who had

a possible bigfoot sighting. Being a bigfoot skeptic she was reluctant to tell me.

However in this case she felt compelled to tell me about her co-worker's encounter because she explained to me that he is a highly intelligent, no-nonsense, straightforward man who would never make up such a story. After a year of missed communication trying to obtain his name and number, I gave up. Fortunately, last week when I spoke to this parent over the phone regarding her son's grades, she mentioned to me that she found his phone number. I immediately called him that evening to set up an interview.

We met Dave at his home on October 9, 2013 at 4:30 pm. Dave has worked as a registered nurse for 7 years and a chiropractor for 23 years which will prove very significant in this unique bigfoot sighting.

Dave saw the creature twice within a two week period. During the first encounter, he was driving home at 11:30 pm. As Dave approached his house, only a few houses down on his street, he observed a large animal on all fours run across the road directly in front of his car. It had dark brown, "rusty-colored" matted hair, a short snout, small pointed ears on the side of its head and it was much larger than a large dog.

The height of its back was as high as a standard kitchen table, the front shoulders were considerably wider than the hips, and it galloped when it ran (bringing the front two legs/arms up together at the same time, then the rear two legs). It crossed the road very quickly. Needless to say, he thought this creature was very odd, especially when he saw it again a few weeks later, walking on two legs!

Dave's second encounter occurred at 6:30 am, just after sunrise. He was inside his home when he looked out his window and noticed a hairy, bi-pedal figure (estimated at 6' 3"), with a very unusual gait, walking down his street. He had a clear, close view of the creature as it walked parallel to his home, only about 30 feet from his windows, although the sunrise was on the direct opposite side of the creature. After watching it from his windows, Dave ran outside to continue viewing this peculiar creature. When he doesn't immediately see it, he walked to the street and eventually saw it about 100 feet from his location, walking away from him on all fours. It turned and looked back at Dave and that's when he realized this was the same creature he witnessed a few weeks earlier that darted in front of his car. It did not have a bulky, muscular build, but instead more of a normal, uniform build for its height.

Based on Dave's years of experience in the medical field, he estimated its weight at about 200 lbs. The arms were slightly longer than a human's and the head was proportionate to its body. The hair was short (maybe only an inch), matted and uniform in length/color throughout.

Chapter Eleven

"Their Watching me"
~Mike Lawrence ~

This encounter took place in March of 2015. It started slowly at first then progressively got more terrifying. Let me give you an image of what my property looks like and my house. My house sits on about a 3/4 acre lot with a large brick house, a warehouse, a bar, and a large carport between warehouse and barn.

My house is all brick with 4 x10 framed roof structures. Over the living room is a two tiered roof due to the vaulted ceiling. Between the two roofs is a row of windows that overlook the living room. The roof is metal with no insulation. In the master bedroom and living room is a widow 8ft up that goes up to the ceiling.

The dining room has two sliding glass doors that are 6' tall by 10' wide; one goes out to the pool, the other goes out to the other back yard. The master bedroom has your standard size sliding glass door going out to the pool.

The backyards are separated by a gate in the middle behind my workshop attached to the house. The pool side is all concrete and bricks with a metal patio roof. The other back yard is all grass with a large maple tree, large oak tree, and another large tree, that is fenced in with a 6' tall fence. The carport, barn and warehouse in the back are all metal roofs. The carport roof is about 40' long 10' deep at a height of 12' sloping down towards the back to about 10'.

The warehouse, barn, and carport are back up to an open field with tall grass, trees, and a game trail that leads down to the seasonal creek. There are motion lights in both sides of the backyard that light up the entire yard. When we first moved in we had now window blinds or curtains in any room or covering any sliding glass door.

This time of year the pool is nothing more than just a pond that frogs have taken refuge in. Every night would be filled with frogs croaking, and crickets chirping. This night was quiet not a single sound, it was an unsettling silence.

I noticed for the first time living here my dogs behavior has changed that night to a frightful behavior. She would not go outside to eat or take care of business before bed. In fact she wouldn't get near any of the sliding glass doors or windows.

That night I heard a loud thud on the carport roof behind the house, which woke me up from my light sleep. I happened to look at the time on my phone and it was 2 am. After a few seconds I heard something bi-pedal run across the roof 4 steps hearing the claws scratch the metal roof as it stepped. On the 4th step I heard it jump then land in the dirt in the backyard. Whatever it was I heard it land with a thud and a couple of steps. My dog heard this too and was trying to crawl under my bed.

Now my dog is a 5 year old chocolate lab mix with her head coming to my hip when she's standing on all fours. She weighs 110lbs, and does not like anyone coming around the house, unless my wife or I bring people in. She is a very good guard dog, and is not afraid of anyone. For me to notice her behavior right now has got me nervous and all my senses are going into overdrive.

As I lay there trying to listen and trying to look out my sliding glass door my motion sensor light turns on and I see a large bipedal dark figure dodge the light going behind my workshop into the shadows.

Then a few seconds later I see the light from the other motion sensor go off from the other backyard down the hallway. The lights turn off and I hear the bipedal creature jump my fence, land, run a few steps, jump and land on the carport roof. It took a couple steps then jumped out into the field and was gone. I tried to gather my thoughts and go back to sleep. My dog then tried crawling out from under my bed and got stuck. At this point my wife asks me what is going on. I tried to explain what I heard and she helped my get dog unstuck. It took me about an hour to finally fall back to sleep. The next day I was trying to contemplate what had happened that night. Thinking it was just a fluke I wrote it off as nothing.

The next night the same thing happens, I hear a loud thud on the carport roof that wakes me up. I look at my clock and it was 2:15am. I hear the same 4 steps with the claws scratching the roof and it jumping then landing in the dirt. This night my dog did not try to crawl under my bed. She didn't do it this time because she got stuck last time. She let out the deepest growl I have ever heard her do, then let out a very deep snarling bark, like don't you dare come near me bark. She started walking backwards into my sons' room away from the sliding door in my room. My wife wakes up and yells at the dog and asks me what's going on.

I tell her I heard it again and she agrees with me now something is going on. I laid there awake for about an hour or so didn't hear anything else, my dog then comes back in and lays down and goes to sleep. Now I feel more relaxed and fall back to sleep. Now this kind of trend goes on for about a week, and I could almost set my alarm to when it came. This happened every morning between 2am and 3am.

My wife was going away for the weekend to her parents' house with our boys. I stayed home with the dog that weekend. Knowing that whatever is happening, it was going to happen again, this time I was alone without worrying about my wife and kids. I wanted to know what this was and I stayed up late that night. My dogs' behavior changed again and she wouldn't get near the sliding doors. I turned off all my lights inside the house and turned on the motion lights.

My screens on the windows are metal screens, not the plastic screens. I hear a scratching noise coming from my window in my bedroom. I walk down my hallway as quietly as I could to see if I can look for what was making the sound. By the time I got there it had stopped. I hear a scratching from the window in my living room. I see a shadow coming in from the street light. Its' arm out stretched scratching my screen looking into my house. I see a large head with pointy ears on top.

It turns to look down the street, and then I see the snout. It turned and went into my carport where my truck is in the darkness. Right after it went into the carport I hear something jump onto my roof of the house over my bedroom. I can't believe there is now more than one. It ran across the roof to the living room and to the windows on the two tier roof. I can feel it looking in staring at me. I leave my living room and try to find my dog. I can't find my dog and she will not come. I eventually find her in my sons' closet laying down in a corner shaking.

I hear the one jump off my roof and then was gone. My dog calmed down which made me calm down. We eventually went to sleep and that was the last time I them for the rest of the year. The shadow in the widow wasn't very big maybe 5' tall probably around 150lbs. I've had encounters before with Dogmen but they were much bigger. My guess is I had 2 juveniles being curious, and messing with me.

Terrifying none the less to see something that doesn't exist, scratching at your window and looking in on you from your skylight windows. I have come to the conclusion they have traveled down the seasonal stream bed and when the stream started drying up, no water to drink, they stopped coming. They stopped as quickly as they started. They came back this year as well about the same time and stopped around the same time.

Chapter Twelve
'' It's tracking me"
Mike (His real name is being protected)

The first encounter January, 2012

During an overnight homeless count (groups of 3 or more search various areas during nighttime hours, collecting data regarding the homeless population), my group and I were on a trail outside of city limits. As we made our way along the trail, we came to a point where we were walking with the river to our right, approximately 10 feet away. Tall, large trees lined the shore. We began to hear a noise coming from the canopy above, that sounded like the trees would shake as we walked, as if something was moving along with us. When we stopped walking, the noise stopped. When we resumed walking, the noise resumed. This continued for several minutes, until we finally heard the noise become louder, and suddenly there was a strange noise, like a muffled grunt, and the tree began whipping back and forth. This was a tree too thick for a human being of any size to be able to whip back and forth. We shined our lights looking for whatever what was doing this, and we were unable to see it. Presumably, it had positioned itself out of sight behind the tree. We immediately left the area.

The first sighting, Spring 2013

I was driving into town on Business Hwy 14 at about 4:00 in the afternoon when the truck in front of me abruptly stopped (as did the vehicle in the oncoming lane), and I saw a large animal, covered in thick, reddish disheveled, matted hair, running on all fours across the road heading east. I did not see the front portion of the animal, but I did see the right buttock and hind leg. The animal moved quite quickly. This occurred within a mile of the location where the first encounter took place.

The picture, July 2013

While hiking on a July afternoon, I came across odd structures in the woods. The structures were obviously newly built. I took pictures of the area. When I got home and looked at the pictures, I realized that a canine like face was hidden in the foliage, watching me take these photos.

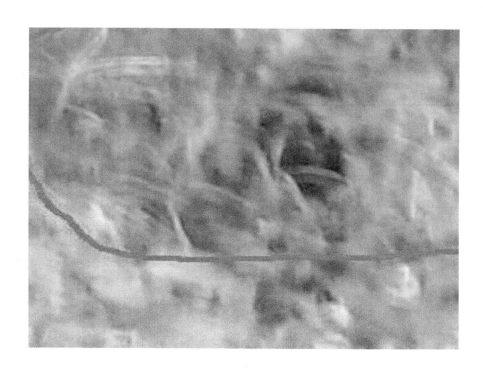

(© By Mike)

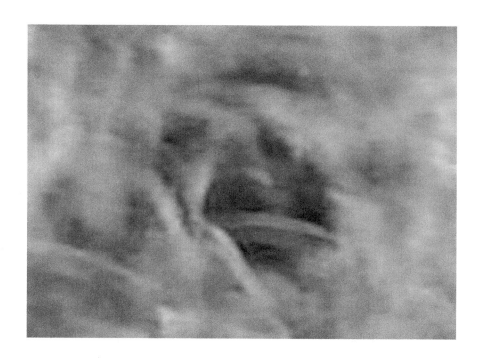

Chapter Thirteen

"Naked" Dogman
~~Dianne Beeson~
NADP Investigator

My coworker had never heard of Dog Man prior to this sighting. On a country road near the water plant, just off Old 32 east near Afton, he was driving to the water tower and spotted something unusual in the light of someone's spotlight on a garage. He backed up the car and looked at it. It was humanoid in shape, in a weird horseshoe pose, giving him the impression that it didn't want to be seen and was eyeing him up to see if he had in fact seen it. It was nude, had dark leathery looking skin, chest hair, short ears, a dog face with a prominent snout. The eyes were not reflecting, but the light was mostly behind it. It remained utterly still, and he watched it for about a minute before driving off. Upon his return about 8 minutes later, it was gone. We drive through there every night about 4 am, and neither of us has seen anything since.

I asked him if it felt evil. He said he didn't really know, but he definitely felt it was "bad news"."

He went on to say it was about man size, but was not a Bigfoot. The witness has seen a Bigfoot before as well, but I don't have that story as yet. But this was NOT a Bigfoot. He also said it had dog legs for the back, but human-like arms."

(Dramatization. © Dianne Beeson NADP Investigator)

Chapter Fourteen
Werewolf of Arkansas
~Vic Cundiff~
~Dogman Encounters~

In October 1972, my husband, our 2 babies, my brother, and I left Leavenworth, KS in our 1968 VW van, on a camping trip to a recreational area in Arkansas called "Beaver Lake." When we finally got there, we found a fairly remote campsite at the far end of the park. We wanted to be alone as the babies woke often during the night and needed to feed. We didn't want to disturb any other campers.

Shortly, after pulling into our campsite, my brother pitched his tent next to the van. The rest of us were going to sleep in the van. The campsite was in an area with a horseshoe-shaped, rocky, terraced ledge, that rose from around 50 feet to around 100 feet as it curved around, behind the 4 campsites. Because of mature trees and thick brush, daylight had trouble poking into our spot.

Fast forward to that night. Some time around 3:30AM, I heard some animal sounds on the ridge that I though were being made by coyotes. The babies were asleep and all was quiet otherwise. I peered out the window, but couldn't see what was making the sounds, because it was so dark. Still hearing odd yips and howls, I laid back down on the back seat. Moments later, there was a huge, crashing, "BANG" on the van wall, right next to my head. My husband leapt up, out of a full sleep. My brother bolted out of his tent and jumped into the van with us. We were all in a panic, looking in every direction, trying to see what had hit the van like that. My brother finally yelled that he saw something moving behind the van. We all turned, just in time to see a large shadow moving about 20 feet behind the van, from left, to right.

After about 20 minutes had passed without any of us seeing movement out there, my husband and brother went out to inspect the van for damage, but found none. We then started hearing pounding steps, coming from the brush, about 50 feet behind us. The guys eased back into the front seat of the van. That's when my husband turned on the headlights and stepped on the brake pedal for rear light. Instantly, there was a huge commotion.

He started the engine and that's when, in the glow from the headlights, we could see a hairy thing, 10 feet away and coming toward the van!

As it got closer, its silver-tipped hairs glistened in the light. It had a greyish streak from its shoulders down its back, to its buttocks. The creature was walking on 2 legs, was around 7 or 8 feet tall, had a barrel-chest, and skinny legs. It never gave us a good view of its eyes, so I can't tell you what they looked like. I could see that the face was not ape-like. It was dog-like. Its ears had tufts of fur on top of them and it was very human-like in its movements and general body structure.

It moved smoothly and quickly, around to the back of the van where it followed the base of the ridge, away from us. That's when it let out a menacing "Huff" and a low, rumbling growl, like a dog. Insanely, my husband and brother bolted from the van, trying to get a better look. That's when a shower of gravel came at us. My husband and brother tore back into the van and burned up the road getting us out of there! I kept looking out the back window and they looked in the rearview mirrors, but none of us ever saw it again. It just didn't seem like a Sasquatch was what we had seen. It seemed too dog-like in its face and was too slim in its body. I still have P.T.S.D. like feelings to this day, due to that encounter!

Pam W SOURCE: Dogmanencounters.com

Eyewitness Drawing ©
Dogman Encounters

Chapter Fifteen
~NADP~
Werewolf Sightings in the UK

Cumberland, ghost of a werewolf?

Something very tall, far taller than their father, nude and grey, something like a man with the head of a wolf—a wolf with white pointed teeth and horrid, light eyes." Mr. Anderson made another and more exhaustive search of the grounds, and discovered, in a cave in the hills immediately behind the house, a number of bones. Amongst them was the skull of a wolf, and lying close beside it a human skeleton, with only the skull missing." - Werwolves by Elliott O'Donnell

The Beast of Ennerdale, 1810

A strange creature that rampaged across the Lake District in north western England in 1810. In five months, it killed almost 300 sheep, often just eating their soft organs and then lapping up their blood.

http://johnknifton.com/2015/09/08/the-beast-of-ennerdale-part-one/

Dogman on road, 1989

We were coming up to a slight right hand bend, there was a six-foot wall with dense foliage on our left and a normal hedgerow bordering a field to our right, that was when I saw something run out in the middle of the road. I wasn't going very fast, probably 30mph maximum and had ample time to slow down so as not to hit it - even to stop but I'm not sure I did. I think I just crawled the car towards it as I couldn't believe what I was seeing.

It had ran out onto the road from my right and had stopped in the middle of the road looking directly at the car. It was covered in thick, mousey grey/brown fur about 2-3 inches long and stood on its rear legs which were jointed with the knee at the rear like a horse or dogs. Its upper arms were down by its side and It held its forearms out in front of it with its hands and long fingers hung down at the wrist.

They may have been long hands or long fingers but I remember the hands from the wrist to the fingers as being very long)

.

It had one foot in front of the other as if in mid-step as it seemed I had caught it by surprise, and while its body was still facing in the direction it was travelling, the head was facing the car.

I remember the eyes which reflected the light back at me as being perfectly round and a green/yellow colour much like that of a cat's eyes when caught in the light. It stood about 5-6 foot tall but was hunched over, even then I'd say it wouldn't have been over 6 foot tall stood erect. It had what seemed to have a very muscular body, not big or bulky, more wiry with powerful limbs, thick chest and small waist. It also had a short tail. maybe 4-5 inches long, definitely not as long as most dogs; personally I'd describe it as more like a goats tail. (After seeing this I mainly searched for sightings of satyrs in the region as this was the closest thing I could relate it to)

Its head and face were like that of a dog or goat, as in not flat like a humans face, but with a muzzle and quite large ears. I do not remember seeing any teeth and it did not have horns. The whole encounter could not have lasted more than a few seconds. I assume it regained its composure after being caught off guard as it took off over the 6 foot wall and through the foliage atop that in a single bound.

https://lindagodfrey.com/2016/04/22/another-u-k-unknown-canine/

Red-eyed dog-headed creature crouched next to highway, 2013

I was driving to pick up my partner from work along the A1231 highway just outside of Sunderland. As I was passing a slip road (exit) I noticed a black figure huddled on the side of the road. I slowed a little as I approached, thinking it was an animal that might dart out in front of the car, but what I saw was like... I'm not quite sure what! The figure had long, shaggy black hair all over its body. It sat in an almost crouched position with its knees drawn up to its chest and its right arm drawn round its legs, resting on its left knee. I slowed down to make sure it wasn't a black rubbish bag. I put my headlights onto full beam, and as I did, it whipped its head around and it was the face of a dog with red eyes !I jumped a little and put my foot down. As I sped by, it watched me go, never blinking.

It creeped me out so much I even locked the doors and rolled the windows up. After I picked up my partner, I drove past the same spot and went a little slower. The figure was gone, but where it had been, the grass was flattened, so something had definitely been there! When I told my partner and our friend about it, my friend called it a "dogman" though she had never heard or seen anything like it herself, certainly not in this area of the world.

http://paranormal.about.com/od/weirdcreaturesmonsters/fl/Dogman-on-the-Highway.htm

Werewolf in Blackpool, 2006

I was walking home down the main road there was only three other people close by and they all saw the same thing.. a very large, grey dog ran from one of the neighbouring streets, it ran into the middle of the road and stopped still. I was close to it and was scared by what I was seeing. The creature reared onto its back legs and howled. It was not like any dog howl I had heard before. It was frightening and high-pitched, almost like a scream, but not a scream. I wanted to run, but I was intrigued. It was at least six feet tall when it was standing and its fangs were three inches long! It looked around at the other three people, then around at me. It growled and took off on all fours back down the street it had come from.

I ran over to the other people who, like me, were mesmerized by what we had witnessed. None of us knew what animal it was. We all had an idea, although no one said anything. The creature we had seen was a werewolf. It couldn't have been anything else. It didn't act like a dog and it didn't even run like a dog; it sort of loped in a weird way.

I didn't do a report about it for the fact that no one would believe me and I would be out of a job for suggesting it I haven't seen the creature since although I have looked regularly and still walk the same way home I even take a camera in case I do see it. I do regularly hear similar noises very late at night. I swear that every word is true even if nobody else admits to seeing it, this is proof that werewolves do exist.

http://paranormal.about.com/library/
blstory_august07_20.htm

Beast of Barmston Drain, 2015

Several reports of "half-man, half-dog" creature. One person claims to have seen the "tall and hairy" beast jump 8ft over a fence, carrying a German Shepherd in its jaw. Mr Covell says another woman described seeing something bounding along the drain on all fours, then stopping and raising up onto its back legs, before running down the embankment towards the water, leaping over to the other side, and vanishing up the embankment and over a wall into some allotments. Another woman described seeing something "half-man, half-dog", along the drain.

"She was terrified, and her dog, which she was walking at the time, began shaking and would not go any further," said Mr Covell.

http://www.hulldailymail.co.uk/Beast-Barmston-Drain-seen-killing-German-Shepherd/story-29234290-detail/story.html

Sighting location

Size chart of the sighting. © NADP

Chapter Sixteen

~NADP~
Beast of the land between the lakes

I wasn't a witness to the fact, just a third person, making observations and having conversations with two individuals who were a part of the incident, who were involved in the whole ordeal. They had just came from the crime scene down in the middle of LBL after being there for over 8 hours. It was around three in the morning and they were taking a much needed reality break.

Two officers of the law. Two grown men who both appeared shaken beyond description. A mixture of fear and confusion, shock and disbelief emanated from them both.

One was paler then the other, a deathly pallor over his skin, and it was this one (I'll name him officer Adam, to protect their identities) that had to sit on the curb of the gas pumps, head between his legs and expel the last bit of his stomach contents.

The other officer (I'll name him officer Bill) came in for some coffee for himself and a cup of water for his partner, then rejoined Adam outside. There were no other customers so I went outside with them to see if I could offer some assistance with the ill man. He gladly took the few Rolaids I had extended in my hand, with his own shaky fingers he struggled to get them into his mouth.

For quite a long while the only thing that was heard were the crickets in the near by fields, the sounds of bugs hitting the fluorescent lights above us hanging from the gas station canopy, and the distant sound of highway traffic that was far and few between as it was in the wee hours of the morning. My mind was buzzing with various scenarios of the cause of their distress....a tragic car accident....possibly a motorcycle wreck...a boating mishap with drown victims....a murder.....a dead body discovered. ('Curiosity killed the cat, but satisfaction brought it back...that's why the cat has nine lives.')

I don't remember sitting down but after about 15 minutes of this hushed stillness I found myself beside them both on the curb staring out at the darkness of the nearby corn pastures, letting my mind paint pictures of imaginary traumas.

Adam spoke first, breaking the silence of obscurity, "I can't believe it...it's not possible...I just can't believe it...". In a hushed agreement, that was almost inaudible, Bill replied, "I know...it was....is....it is so unbelievable...I've never seen anything like this before...", a long pause, a deep breath, and he continued, "... or even heard of anything like this before." I looked at Bill and then at Adam, they were both gazing, open eyed, unblinking, out into the inky color of the night. Adam's bottom lip was trembling slightly, and it wasn't from the slight chill in the late spring air. Something, or some thing had filled them each with a congested fear..

After a few more moments of silent reserve, my patience was rewarded with some slow, fragmented descriptions of their past 8 hours. Bill turned his wide azure blue eyes towards me, they were glazed and blood shot, tired, frightened eyes. With a weary, shaken voice he began to unfold a tale that would forever be embedded within my spirit, like a nasty shadow that lingers around a corner waiting to pounce, to awaken your inner fears once again. Why he decided to tell me of all people was beyond my comprehension, maybe it was an avenue he felt safe to travel upon, to get it off his chest, off his mind.

They were both frequent customers and we knew each other on first name basis, but to divulge such a torrid account of great magnitude, well, I can only say that the fear inside them both at that moment in time had to be released, eased, and extracted from their souls, or else they may have gone mad with unbalanced thoughts.

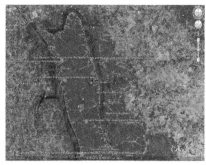

Without interrupting, I sat entranced, listening to every word, absorbing them like an opiate, a spellbinding narcotic that hypnotized me into forgetting the world even existed for the next half hour or so. They had gotten a call to help with an investigation at one of the many rural camp grounds down in LBL. The tourist season was about to start in a few weeks, so as usual there were some early arrivals that had come to claim prime camping spots before the areas were over run with tents, campers and travel trailers The sun was setting low in the sky when they arrived at the scene. Several other official vehicles were already there and there were many more to come as they would soon find out. Many coming from other counties, and a few coming all the way from another state. Several of these to come were coroners from different counties. One coroner vehicle was already present as well as an ambulance, which would prove useless, as there was no one to save. The victims were all dead. Quite dead. Completely, totally and thoroughly deceased. A young married couple that had come down to take it easy for a few days, were the first to discover the ghastly scene. Neither one of them wanted to stay behind while the other went for help, so they both nervously traveled to the nearest town, Grand Rivers, and called the authorities. They did not return to LBL, they merely gave the arriving officer directions to the area of discovery and rented a local hotel room.

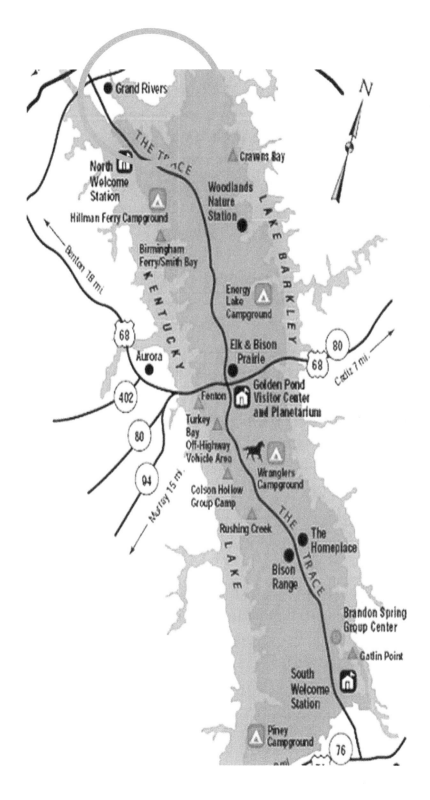

With the sun going down, it got dark pretty fast, so there was a flurry of flood lights from the cruisers being pointed in all directions, along with the excited movements of fifty dollar flash lights being held by nervous, restless hands, searching the trees, the ground, the leaves, the shadows. There was a parked motor home at the site, it's frame being lit by a campfire close by, a fire that had almost went out on it's own, but had been rekindled by the new crowd of men in uniforms so that they could have more light. The front and back doors to the home were open, one of the doors hanging by one hinge in a crooked slant. Through the windows they could see zig zagged movements of luminosity as the beams from flashlights searched the interior. Bloody hand prints slid down the thin metal walls close to the front door and more bloody hand paintings could be seen along the length towards the back door. Their images dancing eerily in the fire light like some ancient tribal symbols .

Adam and Bill did not even want to imagine what was inside the motor home, but then again, they would soon find out, that it wasn't what was 'inside' but what was 'outside' that would change their lives forever. There was already crime scene tape placed in numerous, scattered parts of the area, and little white flags on metal stakes stuck into the ground marking evidence.

Evidence of ripped clothing, bodies and body parts, separated limbs, a pile of bowels, pieces of loose flesh clinging to muscle tissue. What use to be three bodies, that just hours before had been a happy family, on a happy vacation, to create happy memories for years to come; a father, a mother and a young son. The happiness was gone. Destroyed by a psychotic mad man, or was it 'men'? A murderous rage had taken place, one so abhorrently appalling that there were few witnesses to the scene that had kept their composure or held their recently eaten dinners down. At first sight, the victims appeared to be butchered by some un-nameable weapon, possibly an axe, or a chainsaw. Upon further inspection, by the first arriving coroner, the wounds on the bodies were determined not to have been caused by a sharp instrument, but rather by some piercing, well-defined claws, and other wounds by some keen, mordantly long incisors.

Wildcat, bear, wolves? The coroner shook his head in a baffled disagreement with each guess from the officers. The claw marks, for instance, on the back of the fathers corpse were distinctively made by 4 long claws with a smaller digit, like a thumb, on the side, it's span was wider then a man's print, wider and different then a bears mark, with deep deliberate gouges in the flesh.

Artwork by Ginger Bertline

Rake marks from an angry unknown source trying to grab it's prey that was no doubt trying to escape. The wildcat and wolves theory was dismissed as the open wound marks were apparently made by a more grandiose animal source. The bite marks were much larger then any mountain lion, wolf or coyote. Whatever did it had a longer snout, and more sizable teeth. There was also indications in the larger areas of the cadavers, of bite marks where the flesh, meat and bone had been yanked away from the body. Like a human who bites into an apple and leaves the impressions of his bite and teeth marks, so were the open wounds on these individuals. Bears, well, they aren't native to the area, but who knows, maybe a grizzly did sneak in some way, but that was far fetched, he would have had to travel several states and cross several rivers to even get close to that part of Kentucky. Every one present was betting on the 'bear' hypothesis anyway, and no one even thought of anything else to be the cause of such a savage attack. A bear, it had to be a bear.

From the back door of the motor home, an officer stepped down slowly, holding in his hands some type of garment. A dress. A small dress, that would have fit a small girl of around five years old. He informed the on lookers that there were more 'little girls' clothing packed inside the coach. This meant there was a missing person, or an absent body;a member of the family. They all prayed she was still alive somehow, hiding somewhere. A new search began.

As time went by, additional law enforcement employees arrived, as well as a few volunteer rescue squad members. Groups were spread out and assigned areas to examine and explore. Another coroner arrived to assist in the identification and causes of death, and much later a third one showed up, this one from a near by state. All types of samples were placed in plastic bags, marked as evidence, and carefully stowed away. As they were packaging up what appeared to be one of the fathers arms, one of the doctors noticed something wrapped between the dead fingers. Some tweezers slowly untangled a clump of long, grey and brown hairs. This too was placed in a bag, marked and put away to be analyzed at a lab later. From somewhere in the nearby woods, about 50 yards from the campfire, a scream was heard. A mans shriek that turned into a long wail and then to whimpering. As others arrived they could see by the gleam of several flashlights that the cop was holding his hat in one hand and his light in the other. There was blood on his face, the front of his shirt and on the brim of his hat. More blood could be seen dripping on him.

It was coming from above. High in the trees the flash lights swung, searching for the source of the mysterious bleeding. A very small hand could be seen dangling down from a tree limb way up high, as well as a slender lifeless leg that still had a white sock still on the foot.

The missing child had been located. It had been Adam that the blood had trickled upon, hitting his hat first, making him look up, and then feeling the thick cold fluid sprinkling his face then sliding down to his neatly buttoned shirt. It had been Adam that had screamed. The little girl had apparently been carried up the tree and leisurely eaten upon while carefully laid across a large tree branch. More of the same long gray and brown hair was found sticking in the bark of the tree near her body.

After about 7 hours most of the officers were sent away as a new team of investigators arrived. They were told not to talk to anyone of the incident, especially not the media. I am sure that besides Adam and Bill, there were others who had to confess what they saw that night, if in fact this whole event ever really happened. Witness's that had to divulge the awful secret of that atrocious discovery at one of the campgrounds at LBL. About a month after sitting outside with Adam and Bill that night, they stopped in again during one of my midnight shifts. They were both rather quiet, more serious in nature, not like before the incident where they would kid around while drinking their sodas and eating a snack or two. They had both aged in some odd way. Streaks of gray, that had not been there before, highlighted both of their heads of hair. Their faces had lines of worry and showed signs of stress.

I would see them again many times afterwards, but on this particular evening, they informed me that they got word about some of the lab tests that were taken that dreadful night. The tests, on the saliva taken from the bite marks, and from the hair found on the mans fingers and in the tree bark, came back with an unknown species origin. The closest animal that they could be compared to was that of a Canis Lupis, a wolf.

Whether Adam and Bill had played an elaborate hoax on me I'll never really know for sure but their sincerity and fear painted a picture of truth in their eyes and actions. There are several more stories that I have heard about this 'Werewolf' over in LBL that have been told to me over the years after this particular incident. There were several groups of boy scouts that had seen it. Several more campers, fishermen and boaters that had seen it from the safety of of their boats, floating in some of the many bays that touched upon the shoreline. Hikers and bikers have heard its howling and have seen 'something' stalking them while they were on rural trails, hiding amongst the trees and foliage. Hunters have run across deer carcasses that had been brutally torn apart. There was even a pair of curious grave stone rubbers, (those that go out in search of century or more old tombstones then make rubbings by placing paper against the coarse stones and using a piece of charcoal to rub across it thus capturing the images and dates from the stones unto the paper....similar to when as a child you use to take a pencil and rub across a piece of paper on a penny or other coin to see the image of Lincoln or Jefferson.)

that had a fearful encounter with it at one of the old cemeteries. It had actually came up to the car as they were leaving and shook the back end of the vehicle up and down and left terrible scratch marks in the trunk lid as it tried to hold on to the little Toyota while the tires were spinning in the wet grass to get away. These two individuals didn't stop driving until they were about 40 miles away, only then did they dare stop to investigate the damage done. I myself have seen those scratches. Much too wide for any man to have made them. They looked like a heavy metal garden rakes tracks. But you will never read about it in the papers, or hear about it on the news, or get a confession from any law enforcement official or man of office. The media will say it's a bunch of 'Whoo Haa', or just pranks, silly stories, urban legends, lies, tall tales and such. This is tourism country and that means millions of dollars to the area, so you can't scare off business, can you?

Source : Jan Thompson

Artwork by Bart Nunnelly of the LBL Beast

Chapter Seventeen

~NADP~
Werewolf in Erlanger, Kentucky

For about seventy years a tall biped wolf like creature has been making regular appearances in rural Kentucky. This creature was dubbed as the Bearilla. The name Bearilla derives from a description given by the first witness to go public with his sighting, in 1972. While being interviewed by the media, the witness stated the creature looked like a cross between a bear and a gorilla. The last Bearilla sighting occurred in Eastern Kentucky in 2005. A homeowner turned on his porch light, to investigate a strange noise. He was startled to observe what he first thought was child, but after he got a better look, he decide this was no child .... at least not a human one! He described it as being about three feet tall, covered with hair and having a 'dog face". More recently, in the early morning of Good Friday in 2008, something very strange happened in Northern Kentucky...a mysterious sighting of a creature in Erlanger, KY. Many people asked 'could it have been a werewolf?' The Erlanger Werewolf is believed to be around 7 feet tall. Walks upright like a human, but with an obvious hunched back and abnormally long arms. It has a tremendous amount of hair and a very long snout that that of a dog or wolf.

So...are people actually seeing large bi-pedal canines throughout Kentucky? I've received many reports of these cryptids throughout North America and I believe that the phenomenon exists.

Source : Phantoms & Monsters/Witness

Chapter Eighteen
~Vic Cundiff/NADP~
The Black Dog

"What's a Dogman/Black Dog, What I can tell you is that they're commonly reported as being 6, to 7 feet tall, when standing upright on 2 legs. They reportedly have long, canine-looking ears on top of their canine-looking heads. Eyewitnesses have also reported that they smell really bad. They're sometimes seen in groups and are usually seen in remote, forested areas, but not always.

Almost everyone you meet has heard about Bigfoot. Not many realize though, that there's another, much more frightening type of bipedal cryptid stalking the deepest, darkest woods of North America and beyond! Of course, I'm talking about these guys. Dogmen. Have you heard of "The Beast of Bray Road" and/or "The Michigan Dogman?"

Size Charts© NADP Field manual

Size Charts© NADP Field Manual

Size Charts© NADP Field Manual

ACKNOWLEDGMENT

CONTRIPBUTOR
Adam E. Davis
Anthony J. Chaney
David Leidy
Linda Godfrey
Brian Seech
Ronald L. Murphy Jr
Danielle Steadman
Donna Fink
KBRO
Mike Lawrence
Mike"
Dianne Beeson
Vic Cundiff

Sources :
Bart Nunnlley
Phantoms & Monsters
Jan Thompson
NADP

COMING SOON

Germantown Werewolf

Joedy Cook and Adam E. Davis

Made in the USA
Monee, IL
02 December 2021